Ben Stacy Jerrik (Ed.)

Hellikon

Ben Stacy Jerrik (Ed.)

Hellikon

Rheinfelden District, Municipality, Cantons of Switzerland, Aargau

Part Press

Imprint

All parts of this book are extracted from Wikipedia, the free encyclopedia (www.wikipedia.org).

You can get detailed informations about the authors of this collection of articles at the end of this book. The editors (Ed.) of this book are no authors. They have not modified or extended the original texts.

Pictures published in this book can be under different licences than the GNU Free Documentation License. You can get detailed informations about the authors and licences of pictures at the end of this book.

The content of this book was generated collaboratively by volunteers. Please be advised that nothing found here has necessarily been reviewed by people with the expertise required to provide you with complete, accurate or reliable information. Some information in this book maybe misleading or wrong. The Publisher does not guarantee the validity of the information found here. If you need specific advice (f.e. in fields of medical, legal, financial, or risk management questions) please contact a professional who is licensed or knowledgeable in that area.

Cover image: www.ingimage.com
Concerning the licence of the cover image please contact ingimage.

Publisher:
Part Press is a trademark of
International Book Market Service Ltd., 17 Rue Meldrum, Beau Bassin, 1713-01 Mauritius
Email: info@bookmarketservice.com
Website: www.bookmarketservice.com

Published in 2013

Printed in: U.S.A., U.K., Germany. This book was not produced in Mauritius.

ISBN: 978-620-1-58489-1

Contents

References

Hellikon

Hellikon	

Country	Switzerland
Canton	Aargau
District	Rheinfelden

Population	781 (Dec 2012)[1]
- Density	111 /km^2 (287 /sq mi)
Area	km^2 (sq mi)[2]
Elevation	423 m (ft)
Postal code	4316
SFOS number	4251
Localities	Unterdorf, Oberdorf
Surrounded by	Hemiken (BL), Buus (BL), Zuzgen, Obermumpf, Schupfahrt, Weggenstetten
Twin towns	N/A
Website	www.hellikon.ch [3] SFSO statistics [4]

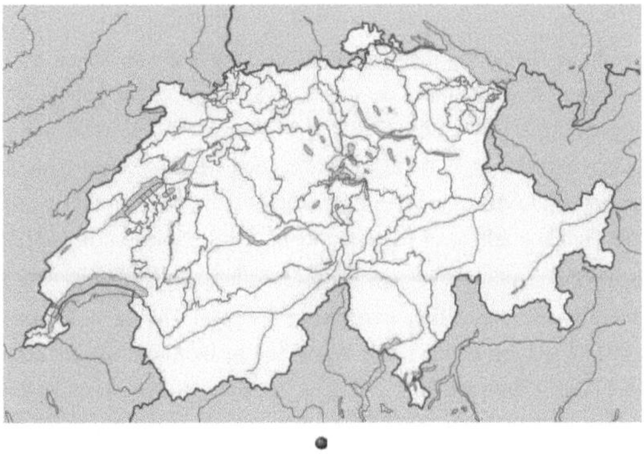

Hellikon

Hellikon (German pronunciation: [ˈhɛlːikχə]) is a municipality in the Rheinfelden district of canton Aargau in Switzerland. It is located in the West of the Fricktal region, around four kilometers Southwest of the border with Germany and has a border with Canton Basel-Landschaft.

Geography

The village lies in the upper reaches of the Möhlin valley (Möhlintal) through which flows the Möhlinbach in a Northwesterly direction towards the Rhine. There are two parts to the village: The lower village (Unterdorf) lies parallel to the Möhlinbach for over one kilometer; The upper village (Oberdorf) lies on the West side of the valley. The flat valley floor alongside the Möhlinbach is, on average, close to 200 meters wide but widens to almost 400 meters when joined by feeder streams. The village is bordered on both sides by the hills of the Tabel Jura (*Tafeljura*) which have very steep lower slopes leading to flat upper plateau's that are intensively farmed.

To the East of the village rises the Wabrig plateau which is over one Kilometer wide and stretches for more than four kilometer between the Lohnberg plateau (above Zuzgen) and the Hersberg (above Wegenstetten). A deeply cut side valley runs in a Westerly direction from the village, separating the Northern Neulig plateau from the Southern Egg plateau.

Hellikon has an area, as of 2009, of 7.05 square kilometers (sq mi). Of this area, 4.29 km^2 (sq mi) or 60.9% is used for agricultural purposes, while 2.3 km^2 (sq mi) or 32.6% is forested. Of the rest of the land, 0.46 km^2 (sq mi) or 6.5% is settled (buildings or roads).[3]

Of the built up area, housing and buildings made up 3.7% and transportation infrastructure made up 2.1%. Out of the forested land, 30.9% of the total land area is heavily forested and 1.7% is covered with orchards or small clusters of trees. Of the agricultural land, 37.0% is used for growing crops and 21.3% is pastures, while 2.6% is used for orchards or vine crops.[3]

The highest point is found on the west side of the Neulig plateau (633 m (ft)) and the lowest at the Möhlinbach (397 m (ft)).

Neighbouring villages in Kanton Aargau are Zuzgen to the Northwest, Obermumpf to the Northeast, Schupfart to the East and Wegenstetten to the Southeast. Neighbouring villages in Canton Baselland are Buus to the West and Hemmiken to the Southwest.

History

The first documented mention of *Hellickonn* dates from 1209.[4] The low jurisdiction was in the possession of the Hofes Zeiningen, which in turn were affiliated to the Stift Säckingen. The gentry and tenants of the "upper juristiction" were Habsburgers. After the Waldshut war of 1468 the whole Fricktal was pledged to the Duchy of Burgandy. The Burgundians were crushingly defeated by the Old Swiss Confederacy during the Burgundian Wars meaning that Hellikon was restoed to Austrian (Habsburg) authority by 1477. Following the country reforms of the Austrian King Maximilian I. Hellikon fell under the authority of "Further Austria" in 1491. The Möhlinbach territory (including Hellikon) came under the administration of the *Kameralherrschaft Rheinfelden*.

During the 17th century, there was very little time when peace prevailed. During the Rappenkrieg, a peasant uprising that lasted from 1612 until 1614 and the Thirty Year's War (1633 to 1638) the villages economic development went into decline. Also during the Pfälzer-Erbfolge Wars (1688 to 1697) foreign troops passed through the region.

As a result of the Treaty of Campo Formio in 1797, the Fricktal became a French protectorate, forming the front line between the French Revolutionary and the Austrian troops in the War of the Second Coalition. On 20 February 1802 Rheinfelden was made a district capital of the newly created Canton of Fricktal, (Principality of Frickgau), joining the Helvetic Republic in August. On 19 February 1803 Canton Fricktal, including Hellikon, was absorbed into Canton Aargau.

At the 1875 Christmas Fair a tragic accident occurred. The stair well of the ten year old school house collapsed, killing 73 people of whom most were children. Between 1850 and 1950 the village population dropped by over twenty percent. Up to the 20th century cereals and grapes contributed the main income, however the phylloxera plague and mildew epidemics effectively ended grape farming in the village. Hellikon transformed from a farming to a commuter community with ever more people moving from the Basel region. Since the beginning of the 1980s the population has grown by almost one third.

In August 2009, Hellikon celebrated its 800th anniversary with a large village fair.

Flag

The blazon of the municipal flags describes "Over blue, a green tripple hill and three yellow morels". The flag was officially adopted in 1924 but it was known much earlier, although never documented. The current form of the flag dates from 1966.[5]

Demographics

Hellikon has a population (as of December 2012) of 781[1] As of June 2009, 5.2% of the population are foreign nationals.[6] Over the last 10 years (1997–2007) the population has changed at a rate of 1%. Most of the population (as of 2000) speaks German(95.6%), with Albanian being second most common (2.4%) and Serbo-Croatian being third (0.5%).[7]

The age distribution, as of 2008, in Hellikon is; 62 children or 8.1% of the population are between 0 and 9 years old and 85 teenagers or 11.1% are between 10 and 19. Of the adult population, 100 people or 13.1% of the population are between 20 and 29 years old. 85 people or 11.1% are between 30 and 39, 134 people or 17.5% are between 40 and 49, and 131 people or 17.1% are between 50 and 59. The senior population distribution is 69 people or 9.0% of the population are between 60 and 69 years old, 54 people or 7.0% are between 70 and 79, there are 41 people or 5.4% who are between 80 and 89,and there are 5 people or 0.7% who are 90 and older.[8]

As of 2000, there were 10 homes with 1 or 2 persons in the household, 108 homes with 3 or 4 persons in the household, and 132 homes with 5 or more persons in the household.[9] As of 2000, there were 253 private households (homes and apartments) in the municipality, and an average of 2.9 persons per household.[7] In 2008 there were 177 single family homes (or 57.7% of the total) out of a total of 307 homes and apartments.[10] There were a total of 2 empty apartments for a 0.7% vacancy rate.[10] As of 2007, the construction rate of new housing units was 6.5 new units per 1000 residents.[7]

In the 2007 federal election the most popular party was the SVP which received 51.3% of the vote. The next three most popular parties were the CVP (20.8%), the Green Party (9.6%) and the SP (7.6%).[7]

The historical population is given in the following table:[11]

Historical population		
Year	**Pop.**	**±%**
1768	252	—
1850	681	+170.2%
1870	705	+3.5%
1900	581	−17.6%
1950	539	−7.2%
2000	747	+38.6%

Religion

From the 2000 census, 389 or 52.1% were Roman Catholic, while 110 or 14.7% belonged to the Swiss Reformed Church. Of the rest of the population, there were 148 individuals (or about 19.81% of the population) who belonged to the Christian Catholic faith. This gives Hellikon the highest proportion of Christ Catholics in Switzerland.[9]

Economy

As of 2007, Hellikon had an unemployment rate of 1.65%. As of 2005, there were 88 people employed in the primary economic sector and about 30 businesses involved in this sector. 47 people are employed in the secondary sector and there are 15 businesses in this sector. 28 people are employed in the tertiary sector, with 11 businesses in this sector.[7]

In 2000 there were 365 workers who lived in the municipality. Of these, 270 or about 74.0% of the residents worked outside Hellikon while 40 people commuted into the municipality for work. There were a total of 135 jobs (of at least 6 hours per week) in the municipality.[12] Of the working population, 14.5% used public transportation to get to work, and 48.8% used a private car.[7]

Small businesses are active in plastic work and building. Most of working population of Hellikon are commuters, working in the larger Fricktal municipalities and the Basel agglomeration.

Transport links

Hellikon lies on the unclassified road running along the Möhlintal from Möhlin to Wegenstetten. A small side road leads to Buus. The nearest intersections with the A3 motorway can be found at Rheinfelden and Eiken. Connection with the public transport network is through the Postauto bus route in either the direction of Möhlin or Weggenstetten.

Sights

The entire village of Hellikon is designated as part of the Inventory of Swiss Heritage Sites.[13]

Education

The entire Swiss population is generally well educated. In Hellikon about 65.1% of the population (between age 25-64) have completed either non-mandatory upper secondary education or additional higher education (either university or a *Fachhochschule*).[7] Of the school age population (in the 2008/2009 school year), there are 59 students attending primary school in the municipality.[9]

The municipality has one Kindergarten and one Primary School. Middle and Secondary schools are located in Möhlin and shared by the entire Möhlintal. Alternative secondary schools are available in Wegenstetten and Zeiningen for those who do not wish to proceed into further education. Thanks to an inter-cantonal agreement, youngsters from the Fricktal may attend the college in Muttenz (canton Baselland).

References

[1] Canton of Aargau Statistical Office (https://www.ag.ch/de/dfr/statistik/statistische_daten/oeffentliche_statistik/bevoelkerung/bevoelkerung_1.jsp), MS Excel document – *Bevölkerungsbestand 2012 nach Gemeinde, Geschlecht u. Nationalität (per 31. Dezember 2012)*] (German) accessed 23 July 2013

[2] Arealstatistik Standard - Gemeindedaten nach 4 Hauptbereichen (http://www.bfs.admin.ch/bfs/portal/de/index/themen/02/03/blank/data/gemeindedaten.html)

[3] Swiss Federal Statistical Office-Land Use Statistics (http://www.bfs.admin.ch/bfs/portal/de/index/themen/02/03/blank/data/gemeindedaten.html) 2009 data (German) accessed 25 March 2010

[4] Bernhard Brogli et al:"Hellikon 1209–2009" Rheinfelden 2009. ISBN 978-3-033-02122-8

[5] Joseph Galliker, Marcel Giger: *Gemeindewappen des Kantons Aargau.* Lehrmittelverlag des Kantons Aargau, Buchs 2004. ISBN 3-906738-07-8

[6] Statistical Department of Canton Aargau -Bereich 01 -Bevölkerung (http://www.ag.ch/staag/) (German) accessed 20 January 2010

[7] Swiss Federal Statistical Office (http://www.bfs.admin.ch/bfs/portal/en/index/regionen/02/key.html) accessed 27-May-2010

[8] Statistical Department of Canton Aargau -Bevölkerungsdaten für den Kanton Aargau und die Gemeinden (Archiv) (http://www.ag.ch/staag/) (German) accessed 20 January 2010

[9] Statistical Department of Canton Aargau - Aargauer Zahlen 2009 (http://www.ag.ch/staag/) (German) accessed 20 January 2010

[10] Statistical Department of Canton Aargau (http://www.ag.ch/staag/) (German) accessed 20 January 2010

[11] *Hellikon* in German (http://www.hls-dhs-dss.ch/textes/d/D1796.php), French (http://www.hls-dhs-dss.ch/textes/f/F1796.php) and Italian (http://www.hls-dhs-dss.ch/textes/i/I1796.php) in the online *Historical Dictionary of Switzerland.*

[12] Statistical Department of Canton Aargau-Bereich 11 Verkehr und Nachrichtenwesen (http://www.ag.ch/staag/) (German) accessed 21 January 2010

[13] ISOS site (http://www.isos.ch/en/nationale/index.asp) accessed 27-May-2010

External links

- Official Website of the Hellikon Municipality (http://www.hellikon.ch)
- *Hellikon* in German (http://www.hls-dhs-dss.ch/textes/d/D1796.php), French (http://www.hls-dhs-dss.ch/textes/f/F1796.php) and Italian (http://www.hls-dhs-dss.ch/textes/i/I1796.php) in the online *Historical Dictionary of Switzerland*.

References

Rheinfelden_District

Rheinfelden District lies in the northwest of the canton of Aargau in Switzerland, in the Fricktal region. Its capital is Rheinfelden. Around 88% of the population live in the conurbation of Basel. There are 14 municipalities, with a population of 45099 (as of 31 December 2012)[1] living in an area of 112.09 km². The population density is around 355 persons per square kilometre.

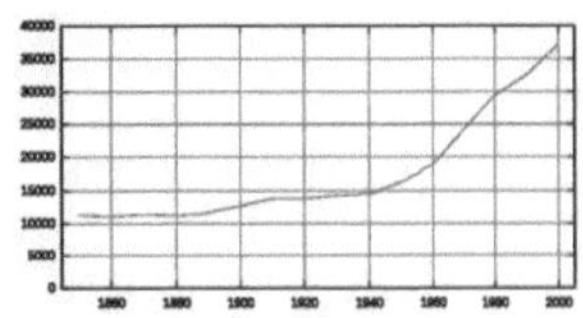

Population

Geography

The Rheinfelden district has an area, as of 2009, of 111.98 square kilometres (sq mi). Of this area, 49.26 km^2 (sq mi) or 44.0% is used for agricultural purposes, while 44.45 km^2 (sq mi) or 39.7% is forested. Of the rest of the land, 14.96 km^2 (sq mi) or 13.4% is settled (buildings or roads).[2]

Demographics

Rheinfelden district has a population (as of December 2012) of 45099.[1] As of June 2009, 21.3% of the population are foreign nationals.[3]

Economy

In 2000 there were 19,081 workers who lived in the district. Of these, 13,553 or about 71.0% of the residents worked outside the Rheinfelden district while 8,272 people commuted into the district for work. There were a total of 13,800 jobs (of at least 6 hours per week) in the district.[4]

Religion

From the 2000 census, 14,982 or 40.2% were Roman Catholic, while 11,153 or 29.9% belonged to the Swiss Reformed Church. Of the rest of the population, there were 2 474 who belonged to the Christian Catholic faith.[2]

Education

Of the school age population (in the 2008/2009 school year), there are 3,090 students attending primary school, there are 1,111 students attending secondary school, there are 734 students attending tertiary or university level schooling, and there are 32 students who are seeking a job after school in the municipality.[2]

Municipalities

Coat of arms	Municipality	Population (31 December 2012)[1]	Area, km²[5]
	Hellikon	781	7.05
	Kaiseraugst	5487	4.91
	Magden	3806	11.01
	Möhlin	10501	18.79
	Mumpf	1290	3.1
	Obermumpf	998	5.03
	Olsberg	355	4.62
	Rheinfelden	12182	16.03
	Schupfart	759	7.04
	Stein	2975	2.81
	Wallbach	1846	4.51
	Wegenstetten	1038	7.13
	Zeiningen	2229	11.37
	Zuzgen	852	8.39

References

[1] Canton of Aargau Statistical Office (https://www.ag.ch/de/dfr/statistik/statistische_daten/oeffentliche_statistik/bevoelkerung/bevoelkerung_1.jsp), MS Excel document – *Bevölkerungsbestand 2012 nach Gemeinde, Geschlecht u. Nationalität (per 31. Dezember 2012)*] (German) accessed 23 July 2013

[2] Statistical Department of Canton Aargau - Aargauer Zahlen 2009 (http://www.ag.ch/staag/) (German) accessed 20 January 2010

[3] Statistical Department of Canton Aargau -Bereich 01 -Bevölkerung (http://www.ag.ch/staag/) (German) accessed 20 January 2010

[4] Statistical Department of Canton Aargau-Bereich 11 Verkehr und Nachrichtenwesen (http://www.ag.ch/staag/) (German) accessed 21 January 2010

[5] Swiss Federal Statistical Office-Land Use Statistics (http://www.bfs.admin.ch/bfs/portal/de/index/themen/02/03/blank/data/gemeindedaten.html) 2009 data (German) accessed 25 March 2010

Municipality

WARNING: Article could not be rendered - ouputting plain text.
Potential causes of the problem are: (a) a bug in the pdf-writer software (b) problematic Mediawiki markup (c) table is too wide

The Ponce City Hall, in the city of Ponce, Puerto RicoPonce, is the seat of the government for both the city and the surrounding barrios making up the municipality. A municipality is usually an urban administrative division having corporationcorporate status and usually powers of self-government or jurisdiction. The term municipality is also used to mean the governing body of a municipality. A municipality is a general-purpose administrative subdivision, as opposed to a special-purpose district. The term is derived from French "municipalité" and Latin "municipalis".Etymology The English languageEnglish word "Municipality" derives from the Latin social contract "Municipium", meaning duty holders, referring to the Latin communities that supplied Rome with troops in exchange for their own incorporation into the Roman state (granting Roman citizenship to the inhabitants) while permitting the communities to retain their own local governments (a limited autonomy).Generic term The term "municipality" is a generic term, and can describe any political jurisdiction from a sovereign state, such as the MonacoPrincipality of Monaco, or a small village, such as West Hampton Dunes, New York. The territory over which a municipality has jurisdiction may encompass only one populated place such as a city, town, or village several of such places (e.g., early jurisdictions in the state of New Jersey (1798-1899) as Township (New Jersey)townships governing several villages, Municipalities of Mexico) only parts of such places, sometimes boroughs of a city such as the 34 municipalities of Santiago, Chile.In a number of countries such as Finland, all land must belong to a municipality. In contrast, in other countries the municipality covers only a single populated place and the rest - unincorporated territory, which may have substantial population such as villages - is governed directly by the county. Municipalities may be subdivided into several types or tiers, such as cities, towns, or rural municipalities.Political powers The power of municipalities range from virtual autonomy to complete subordination to the State (polity)state. Municipalities may have the right to tax individuals and corporations with income tax, property tax, and corporate income tax, but may also receive substantial funding from the state. In various countries In various countries, a municipality is the smallest administrative subdivision to have democratically elected representation. Municipalities are sometimes referred to as "communes" (for example, Communes of FranceFrench commune, Spanish ayuntamiento, Italian comune, Romanian comună, Swedish kommun Norwegian/Danish kommune and German Kommune).English speakingIn Australia, the term local government area (LGA) is used in place of the generic municipality. Here, the "LGA Structure covers only incorporated areas of Australia. Incorporated areas are legally designated parts of states and territories over which incorporated local governing bodies have responsibility."In Canada, municipalities are local governments established through provincial and territorial legislation, usually within general municipal statutes. Types of municipalities within Canada include citycities, District municipalitydistrict municipalities, municipal districts, municipalities, Parish (administrative division)parishes, Rural municipalityrural municipalities, towns, townships, villes, and villages among others. The Province of Ontario has different tiers of municipalities, including lower, upper, and single tiers. Types of upper tier municipalities in Ontario include counties and regional municipalities. Nova Scotia also has regional municipalities, which include cities, counties, districts, or towns as municipal units.In the Local government in the United KingdomUnited Kingdom, the term was used until 1974 in England and Wales, and until 1975 in Scotland and 1976 in Northern Ireland, "both for a city or town which is organized for self-government under a municipal corporation, and also for the governing body itself. Such a corporation in Great Britain consists of a head as a mayor or provost, and of superior members, as aldermen and councillors". Since local government reorganisation, the unit in England, Northern Ireland and Wales is known as a district, and in Scotland as a council area. A district may be awarded borough or city status, or can retain its district title.In the United States, "municipality" is usually understood as a city, town, village, or other local government unit, formed by municipal charter from the state as a municipal corporation. In a state law context, some U.S. state codes define "municipality" more widely, from the state itself to any political subdivisions given jurisdiction over an area that may include multiple populated places and unpopulated places.In Jersey, a municipality refers to the honorary officials elected to run each of the 12 Parishes of JerseyParishes into which it is subdivided. This is the highest level of regional government in this jurisdiction. Chinese speakingIn the People's Republic of China, a direct-controlled municipality (in pinyin: zhíxiáshì) is a city with equal status to a Province (China)province: Beijing MunicipalityBeijing, Tianjin MunicipalityTianjin, Shanghai MunicipalityShanghai, and Chongqing MunicipalityChongqing (see Municipality of China) In Republic of ChinaTaiwan (Republic of China), a special municipality (in pinyin: Political divisions of Taiwanzhíxiáshì) is a city with equal status to a province: Kaohsiung, New Taipei, Taichung, Tainan and Taipei. (see Special municipality of Taiwan) Municipalities by country Country Term Example Subdivision of Quantity Notes Further reading Albaniabashki (urban) or komunë (rural)Municipalities of AlbaniaAlgeriabaladiyahdaïra (district), which is part of a Wilayas of Algeriawilaya (province) 1,541 Municipalities of AlgeriaArgentinamunicipalidad is the institution in charge of local administration. The provinces organize the municipalities for each city in their territories according to their own municipiomunicipal regime. AustriaGemeindedistrict#Austriadistrict (Bezirk), which is in turn part of a States of Austriastate (Bundesland).Municipalities of AustriaBangladeshPouroshabha / PouroshavaUpazilas of Bangladeshupazila or subdistrict, which is in turn part of a Districts of Bangladeshdistrict 311 Municipalities of BangladeshBelgiumgemeente/commune a province (Provinces of regions in Belgiumprovincie/province) or of the Brussels-Capital RegionMunicipalities in BelgiumBoliviamunicipioArani Municipalityprovince, which is part of a Department (country subdivision)departamento. 327 (2005) Municipalities of BoliviaBosnia and Herzegovinaopćina or opština in Federation of Bosnia and Herzegovina : of a canton (Cantons of Bosnia and Herzegovinakanton). In Republika Srpska, municipality is first order of government after entity government.Municipalities of Republika SrpskaBrazilMunicipalities of BrazilmunicípiosSão Paulo (largest), Borá (smallest) states (States of Brazilestados) 5,564 is the smallest political-administrative division; there is no equivalent to a county level. A city (cidade) is defined in Brazilian law as the urban seat of a municipality, and a municipality always has the same name as the corresponding city. Brazilian law establishes no difference between cities and towns; all it takes for an urban settlement to be called a "city" is to be the seat of a municipality, and some are very small. Other settlements have no form of local government and are under the authority of the municipality they are in, although in some cases the municipal government may set up local administrative offices there. The Federal District (Brazil)Federal District (the area of the national capital city, Brasília) has special status and is not divided into municipalities. The Fernando de Noronha and Saint Peter and Saint Paul ArchipelagoSt. Peter and St. Paul archipelagos together comprise a unique case of a "state district" under the direct administration of the state government of Pernambuco. Apart from these exceptions, all land in Brazil, even the remotest wilderness areas, belongs to the territory of some municipality, and, hence, technically under the jurisdiction of a "city." No point in the country lies in a non-incorporated area, and that explains why some municipalities in sparsely populated areas such as the Amazon region can be larger than many sovereign countries.Municipalities of BrazilBulgariaBulgarian languageBulgarian: община, obshtinaAksakovo Municipality an oblast 264 Municipalities of BulgariaKingdom of Cambodia () refers to city hall Example, if they say Phnom Penh Municipality it means Phnom Penh City Hall, but sometimes a municipality refers to the capital city, which is Phnom PenhPhnom Penh City. Chilemunicipalidad provinces is a legal entity which administers one or more comunacommunes (comuna) which are the third-level division of the country. The first division are Regions of Chileregions which a next divided into Provinces of Chileprovinces (provincia). These provinces are next divided into comunas which are assigned to a municipality for administration. In almost most cases the municipality and the comuna have the same name, but the constitution permits a single municipality to be responsible for more than one commune.Colombiamunicipio a Departments of Colombiadepartment (departamento) It is also subdivided into Corregimientos and Veredas. Municipalities of ColombiaCosta RicaMunicipalidad Province (Provinces of Costa Rica)Cantons of Costa RicaCantones of Costa RicaCroatiaopćina county (Counties of Croatiažupanija)Municipalities of CroatiaCzech Republicobec a kraj (kraj)DenmarkkommuneAabenraa Municipality a Regions of Denmarkregion Counties (Counties of Denmarkamter) were abandoned in Denmark on January 1, 2007.Municipalities of DenmarkDominican RepublicmunicipioProvinces of the Dominican RepublicprovinceMunicipalities of the Dominican RepublicEstoniaomavalitsus Are vald (Are Parish) Counties of Estoniacounty (maakond) 226 is the smallest division of local government. Municipalities can be both towns (linn) and parishes (vald).Municipalities of EstoniaFinlandkunta is independent and constitutes local government; each co-operates with municipalities nearby in a sub-region (seutukunta) and region (maakunta). A municipality can freely call itself a "city" (kaupunki).Municipalities of FinlandFranceCommune in Francecommune department (département in Francedépartement) which is part of a region (région)36,600 GermanyGemeinde district (Districts of GermanyKreis). Larger entities of the same level are called towns (Stadt). In less populated regions, municipalities are often put together into collective municipalities (Verbandsgemeinde) 12013 Municipalities of GermanyGreecedemos (δήμος, pl. δήμοι) or kinotita (κοινότητα, pl. κοινότητες) with lesser population a Municipalities and communities of Greecemunicipality is either, which are then part of a prefecture (Prefectures of Greecenomos, νομός) and then a larger region known as a peripheriesperiphery (περιφέρεια, pl. περιφέρειες). Municipalities are third-level administrative divisions and their heads (mayors in demi, presidents in kinotites) are appointed via popular vote held every four years.Greenland kommunea Kujalleq is the first-level administrative division of country. Municipalities of GreenlandHaiticommune an arrondissement, which is part of a department (département).Hondurasmunicipio was in 1895 originally the subdivision of the district (distrito), but as districts have fallen into disuse, it is now the subdivision of the department (departamento). 298 Municipalities of HondurasHungarytelepülési önkormányzat a county (megye). 3,168 (in 2005) IcelandsveitarfélagReykjavíkRegions of IcelandRegions 74 Municipalities in Iceland are the only self-governing local administrative units. Municipalities of IcelandIndianagar palika A municipality is often referred to as a town. It is neither a village nor a big city. Usually, a municipality would have 20,000 or more people, but if it exceeds 500,000 it becomes a municipal corporation (Nagar nigam). Ward councillors of municipal corporations are nominated by the subdivisional officers and elected by the people. The chairman is also a nominee of the subdivisional officer who is a government officer in India.Municipalities of IndiaIsrael a municipality generally takes one of three forms: city council (Israel)city councils, which govern a large municipality, local council (Israel)local councils, which govern a small municipality, and regional council (Israel)regional councils, which govern a group of communities, often but not necessarily of a rural nature. Italycomune a province (Provinces of Italyprovincia) which is part of a region (Regions of Italyregione). The term "municipality" is reserved for subdivisions of larger comuni (in particular, the comune of Rome).Japan a municipality is the sphere of government within the prefectures, the sub-division of the state. Municipalities of Japan include Cities of Japanshi (cities), List of towns in Japancho (towns), and Villages of Japanson (villages). Kenya a municipality is one of four types of local authorities. Nearly 50 major towns are given the municipality status. Latviasing.:novads, plur.:novadi) Aglona novads (Aglona Municipality) A municipality normally consists of Amalgamation (politics)amalgamated Civil parishparishes (sing.:pagasts, plur.:pagasti).Municipalities of LatviaLebanon a Districts of Lebanondistrict (Arabic languageArabic: Qadaa) which is part of a Governorate (Region or Province, Arabic: Mouhafazah).Libya the municipality level is that of the Basic People's Congress (country subdivision)Basic People's Congress. Large cities are subdivided. Municipalities of LibyaLithuaniasavivaldybė Kalvarijos savivaldybė (Kalvarija Municipality) a district (apskritis) and is subdivided into elderates (seniūnija).Municipalities of LithuaniaLuxembourgcommunes of Luxembourgcommunes are the lowest divisions Republic of MacedoniaMacedoniaopštinaAračinovo Municipality the country (top-level) or the City of Skopje 84 (2004) current set established in 2004, reduced from 123 created in 1996. Municipalities of the Republic of MacedoniaMexicomunicipio / delegaciónAsientos Municipality a state (States of Mexicoestado) or the DF a borough (delegación) is a subdivision of the Federal District (Mexico)Federal DistrictMunicipalities of Mexico and boroughs of the Mexican Federal DistrictMoldovamunicipiuMunicipiului Chişinău is a main city or town and surrounding villages ranked by law at this level. Moldova#Administrative_divisionsMontenegroopštinaAndrijevica Municipality the country (top-level) Municipalities of MontenegroNepal Nagarpalika 58 (after the census in 2001)+33 new municipalities were in 2012 a municipality is a town not large enough in population or infrastructure to qualify as a sub-metropolitan areametropolitan city. Netherlandsgemeente a province (Provinces of the Netherlandsprovincie) 408 + 3 special municipalitiesMunicipalities in the NetherlandsNew Zealand a municipality is part of either a "city" (mostly urban) or a "district" (mostly rural). The term "municipality" has become rare in New Zealand since about 1979 and has no legal status. (See territorial authorities of New Zealand.) Nicaraguamunicipio a department (departamento) or of one of the two Autonomous Regions, Región Autónoma del Atlántico Norte and Región Autónoma del Atlántico Sur.Municipalities of NicaraguaNorwaykommune a county (Counties of Norwayfylke) 430 (2009) Municipalities of NorwayPalestinian National Authority 105 municipalities are localities with populations over 4,000 and have 13-15 council members. Municipality (Palestinian Authority)Pakistan of a tehsil/taluka or sub-district of a "Districts of Pakistandistrict". Local government in PakistanParaguaymunicipalidad a Department (country subdivision)departament (departamento)Perumunicipio a Provinces of Peruprovince (provincia), which is part of a Departments of Perudepartment (departamento). As of 2002 a department is now called a region (región). is another term for Districts of Perudistrict (distrito) and is the lower-level administrative subdivision.Philippinesmunicipalities of the Philippinesbayan or munisipyo is a town with a popularly elected administration including a mayor and is part of a province (Provinces of the Philippineslalawigan or probinsya) (except for the independent municipality of Pateros, Metro Manila in the Metro ManilaNational Capital Region). Municipalities are not the smallest division of local government and are composed of barangays.PolandgminaGmina Annopol county (powiat).PortugalmunicípioAveiro Municipality municipality is a directly elected local area authority generally consisting of a main city or town and surrounding villages, with wide-ranging local administration powers. Apart from the municipality of Corvo (Azores)Corvo, however, municípios are not the smallest administrative unit in Portugal, that being the freguesia (civil parish). For central government purposes, Portuguese municipalities are grouped into districts (Districts of Portugaldistritos). In Portugal, a municipality (Municipalities of Portugalmunicípio/Concelhos of Portugalconcelho) is the primary local administrative unit. Although part of a district (Districts of Portugaldistrito) for certain national administrative purposes, the municipality is not subordinate to the district and decentralization is doing away with the districts. A municipality contains one or more freguesias.Municipalities of PortugalPuerto Ricomunicipio municipality consists of an urban area (termed a city or town) plus all of its surrounding barrios comprising the municipality. It has a popularly elected administration and a municipal mayor. The seat of the municipal government is located in such urban area and serves the entire municipal jurisdiction. Ayuda: Divulgación de Resultados de la Comisión Estatal de Elecciones - Municipio. Comision Estatal de Elecciones. Estado Libre Asociado de Puerto Rico. San Juan, Puerto Rico. 2004. Retrieved 6 December 2011. Autonomous Municipalities Law. (in Spanish) Retrieved 24 October 2011.Municipalities of Puerto RicoRomaniamunicipiu is a town or a city ranked by law at this level. A communes of Romaniacommune is the lowest subdivision of a judeţ. Municipalities in RomaniaRussia several types of municipalities ("municipal formations") exist; see subdivisions of Russia#Municipal divisionssubdivisions of RussiaSan Marino there are also eight minor municipalities, castelli. Serbiaopština a district (okrug)Slovakiaobecokresdistrict (okres)List of villages and municipalities in Slovakia2,891Sloveniaobčinaregion (regija) 212 a municipality (Municipalities of Sloveniaobčina) is the primary local administrative unit. There are 212 of them, 11 of which have a special "Urban" status with additional autonomy.Municipalities of SloveniaSouth AfricaDistrict municipality (South Africa)district municipalities and Metropolitan municipality (South Africa)metropolitan municipalities are subdivisions of the Provinces of South Africaprovinces, and Local municipality (South Africa)local municipalities are subdivisions of district municipalities. Spainmunicipio a municipality (municipio) is the primary local administrative unit. It is a part of a province (Provinces of Spainprovincia) for all national administrative purposes. In the Galicia (Spain)Galicia region, the municipalities are called concellos, and in the Principality of Asturias region, a municipality is called a conceyu. In these two regions a municipality contains one or more parroquias. Municipalities of SpainSwedenkommunAle Municipality 290 is self-governing according to the Swedish constitution and constitutes local government. Before 1971, a municipality could be called a town (stad), a köping or a rural municipality (landskommun); present municipalities which used to be towns are still commonly called towns. Sweden is also divided into 25 counties (Counties of Swedenlän), each containing one or more of the municipalities.Municipalities of SwedenSwitzerlandcommune/Gemeinde/comune a canton (Cantons of Switzerlandcanton/Kanton/cantone) and defined by cantonal law.Municipalities in SwitzerlandThailandthesaban None 1,602 There are three levels of municipalities, and municipalities are not self-governed except for Bangkok and Pattaya. Municipalities introduced in 1935, Bangkok became self-governed in 1972, and Pattaya became self-governed in 1978. ThesabanUkraine there are two types of municipalities (or rada — council): urban and rural. Both are types of lower administrative division in the country. Urban municipalities could be city or settlement (town). Cities always carry special status: national (two), oblast (at least one in each region), and raion (the rest). Urban municipalities often consist of several other smaller ones. Bigger cities are divided into number of city raions (districts). Rural municipalities are much smaller and may combine several villages and rural settlements (smaller villages). Two cities Kiev and Sevastopol are organized as special municipalities that are independent from their regional government (oblast / republic) within they are located. United Arab Emirates an emirate is defined by the law of the specific emirates. Turkey belediye is a local government authority and there are two types of municipalities: metropolitan (büyükşehir) municipality, district (alt/ilçe) municipality. UruguaymunicipioMunicipalities of UruguayVenezuelamunicipioAnaco Municipality a state, as well as a subdivision of the Venezuelan Capital DistrictCapital District (Subdivisions of Venezuelaestado)Municipalities of VenezuelaVietnamDirect-controlled municipalityHo Chi Minh City (largest) 5 In the Vietnam, a direct-controlled municipality (Thành phố trực thuộc trung ương in Vietnamese) is a city with equal status to a Provinces of Vietnamprovince: Hanoi, Ho Chi Minh City, Hai Phong, Da Nang and Can ThoMunicipalities of VietnamIn Portuguese language usage, there are two words to distinguish the territory and the administrative organ. When referring to the territory, the word concelho is used, when referring to the Organs of stateorgan of State, the word município is used. This differentiation is still in use in Portugal and some of its former overseas provinces, but it's no longer in use in Brazil.See alsoCouncil of European Municipalities and RegionsCouncil-manager governmentCreature of statuteLists of municipalitiesMayor-council governmentMunicipal corporationMunicipal governmentReferences External links

Cantons_of_Switzerland

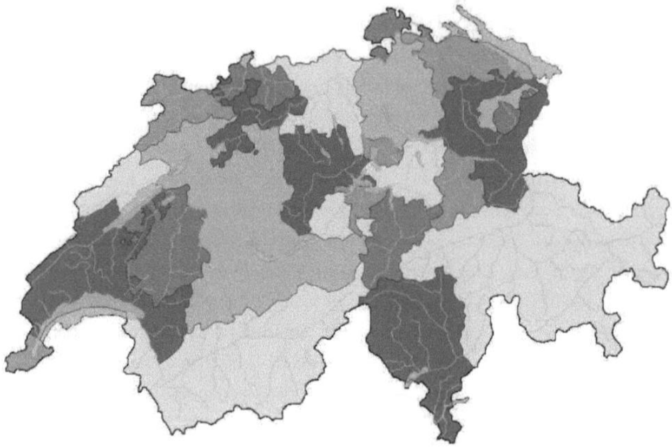

Valais

Ticino

Graubünden (Grisons)

Geneva

Vaud

Neuchâtel

Jura

Bern

Thurgau

Zurich

Aargau

Lucerne

Solothurn

Basel-Land

Schaffhausen

Uri

Schwyz

Glarus

St. Gallen

AI

AR

Obwalden

Nidwalden

Zug

Fribourg

Basel-Stadt

France

Italy

Liecht.

Austria

Germany

The 26 **cantons of Switzerland** are the member states of the federal state of Switzerland. Each canton was a fully sovereign state[1] with its own border controls, army and currency from the Treaty of Westphalia (1648) until the establishment of the Swiss federal state in 1848. The most recently created canton is the Canton of Jura, which separated from the Canton of Bern in 1979.[2]

The name is derived from the French language word *canton* meaning *corner* or *district* (from which the term Cantonment is also derived).

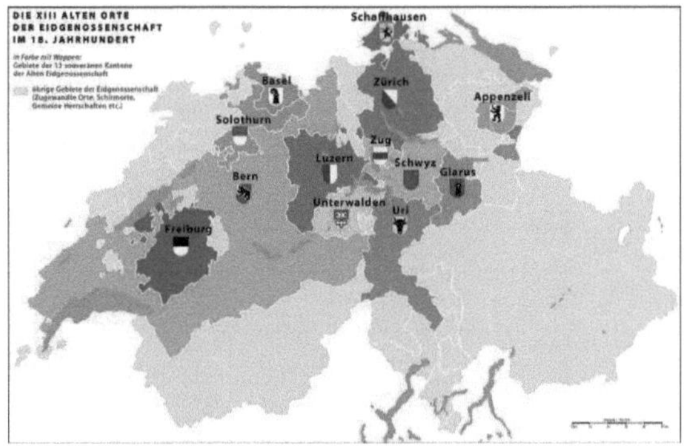

The 13 cantons of the Old Swiss Confederacy (1513-1798).

History

In the 16th century, the Old Swiss Confederacy was composed of 13 sovereign cantons, and there were two different kinds: six land (or forest) cantons and seven city (or urban) cantons. Though they were technically part of the Holy Roman Empire, they had become *de facto* independent when the Swiss defeated Emperor Maximillian in 1499.[] The six forest cantons were democratic republics, whereas the seven urban cantons were oligarchic republics controlled by noble families.

Constitution

Each canton has its own constitution, legislature, government and courts.[3] Most of the cantons' legislatures are unicameral parliaments, their size varying between 58 and 200 seats. A few legislatures are general assemblies known as *Landsgemeinden*. The cantonal governments consist of either five or seven members, depending on the canton.[4] For the names of the institutions, see List of legislative and executive councils of the Cantons of Switzerland.

The Swiss Federal Constitution declares the cantons to be sovereign to the extent their sovereignty is not limited by federal law.[3] The cantons also retain all powers and competencies not delegated to the Confederation by the Constitution. Most significantly, the cantons are responsible for healthcare, welfare, law enforcement and public education; they also retain the power of taxation. The cantonal constitutions determine the degree of autonomy accorded to the municipalities, which varies but almost always includes the power to levy taxes and pass municipal laws. The sizes of the cantons vary from 37 km² to 7,105 km²; the populations vary from 15,471 to 1,244,400.

Direct democracy

As on the federal level, all cantons provide for (half-) direct democracy. Citizens may demand a popular vote to amend the cantonal constitution or laws, or to veto laws or spending bills passed by the parliament. General popular assemblies (*Landsgemeinde*) are now limited to the cantons of Appenzell Innerrhoden and Glarus. In all other cantons democratic rights are exercised by secret ballot.

List

The cantons are listed in the order given in the federal constitution.[a]

Coat of arms	Abbr	Canton	Since	Capital	Population[1]	Area (km²) [2]	Density (per km²) [3]	No. munic.[4]	Official languages
	ZH	**Zurich**	1351	Zurich	1406083		701	171	German
	BE	**Bern**	1353	Bern	985046		158	383	German, French
	LU	**Lucerne**	1332	Lucerne	381966		233	87	German
	UR	**Uri**	1291[5]	Altdorf	35382		33	20	German
	SZ	**Schwyz**	1291[5]	Schwyz	147904		143	30	German
	OW	**Obwalden**	1291[5] or 1315 (as part of Unterwalden)	Sarnen	35878		66	7	German
	NW	**Nidwalden**	1291[5] (as Unterwalden)	Stans	41311		138	11	German
	GL	**Glarus**	1352	Glarus	39217		51	3	German
	ZG	**Zug**	1352	Zug	116559		416	11	German
	FR	**Fribourg**	1481	Fribourg	284668		141	167	French, German
	SO	**Solothurn**	1481	Solothurn	259836		308	122	German
	BS	**Basel-Stadt**	1501 (as Basel until 1833/1999)	Basel	194090		5,072	3	German
	BL	**Basel-Landschaft**	1501/1833[6]	Liestal	277973		502	86	German
	SH	**Schaffhausen**	1501	Schaffhausen	77139		246	27	German
	AR	**Appenzell Ausserrhoden**	1513 [7]	Herisau[5]	53313		220	20	German
	AI	**Appenzell Innerrhoden**	1513[7]	Appenzell	15789		87	6	German
	SG	**St. Gallen**	1803[8]	St. Gallen	483101		222	85	German
	GR	**Graubünden**	1803[9]	Chur	193388		26	180	German, Romansh, Italian
	AG	**Aargau**	1803	Aarau	627893		388	220	German

	TG	**Thurgau**	1803[10]	Frauenfeld[6]	254528		229	80	German
	TI	**Ticino**	1803[11]	Bellinzona	336943		110	157	Italian
	VD	**Vaud**	1803[12]	Lausanne	725944		188	339	French
	VS	**Valais**	1815[13]	Sion	317022		53	143	French, German
	NE	**Neuchâtel**	1815/1857[14]	Neuchâtel	173183		206	53	French
	GE	**Geneva**	1815	Geneva	472530		1,442	45	French
	JU	**Jura**	1979[15]	Delémont	70542		82	64	French
	CH	**Switzerland**		Bern	7,968,705		174	2,596	German, French, Italian, Romansh

The two-letter abbreviations for Swiss cantons are widely used, e.g. on car license plates. They are also used in the ISO 3166-2 codes of Switzerland with the prefix "CH-" (*Confœderatio Helvetica*, Switzerland), e.g. CH-SZ for the canton of Schwyz.

Half-cantons

Six of the 26 cantons are traditionally, but no longer officially, called "half-cantons" (German: *Halbkanton*, French: *demi-canton*, Italian: *semicantone*), reflecting a history of mutual association or partition.

The half-cantons are identified in the first article of the Swiss Federal Constitution of 1999 by being joined to their other "half" with the conjunction "and":

> The People and the Cantons of Zurich, Bern, Lucerne, Uri, Schwyz, Obwalden and Nidwalden, Glarus, Zug, Fribourg, Solothurn, Basel-Stadt and Basel-Landschaft, Schaffhausen, Appenzell Ausserrhoden and Appenzell Innerrhoden, St. Gallen, Graubünden, Aargau, Thurgau, Ticino, Vaud, Valais, Neuchâtel, Geneva, and Jura form the Swiss Confederation.
>
> —Article 1 of the Federal Constitution of the Swiss Confederation (underlining not in original)[16]

The 1999 constitutional revision retained this distinction, on the request of the six cantonal governments, as a way to mark the historic association of the half-cantons to each other.[17] In contrast, the first article of the 1848 and 1874 constitutions constituted the Confederation as the union of "twenty-two sovereign cantons",[18] referring to the half-cantons as "Unterwalden (above and beneath the woods)", "Basel (city and country)" and "Appenzell (both Rhoden)".[19] While the older constitutions referred to these states as "half-cantons", a term that remains in popular use, the 1999 revision and official terminology since then use the appellation "cantons with half a cantonal vote".[20]

With their mutual association a purely historical matter, the half-cantons are since 1848 equal to the other cantons in all but two respects:[21]

- They elect only one member of the Council of States instead of two (Cst. art. 150 par. 2).
- In popular referendums about constitutional amendments, which require for adoption a national popular majority as well as the assent of a majority of the cantons (*Ständemehr / majorité des cantons*), the result of the half-cantons' popular vote counts only one half of that of the other cantons (Cst. arts. 140, 142). This means that for purposes of a constitutional referendum, at least 12 out of a total of 23 cantonal popular votes must support the amendment.[22]

The reasons for the association between the three pairs of half-cantons are varied:

Caricature of the division of Basel, 1833

- *Unterwalden* never consisted of a single unified jurisdiction. Originally, Obwalden, Nidwalden, and the Abbey of Engelberg formed distinct communities. The collective term *Unterwalden* remains in use, however, for the area that partook in the creation of the original Swiss confederation in 1291 with Uri and Schwyz. The Federal Charter of 1291 called for representatives from each of the three "areas".[23] [24]
- The canton of Appenzell divided itself into an "inner" and "outer" half ("Rhoden") as a consequence of the Reformation in Switzerland in 1597:[25] Appenzell Innerrhoden (Catholic) and Appenzell Ausserrhoden (Protestant).
- The canton of Basel divided itself as a consequence of a revolt of the Basel countryside in 1833, in order to promote equality among its citizenry, combating claims between rural and city residents over preferential status:[26] Basel-Landschaft and Basel-Stadt.

Names in national languages

(Names appear in bold when corresponding to the cantonal official language)

Abbr	Common English	Other English forms	German	French	Italian	Romansh
AG	Aargau	Argovia	**Aargau**	Argovie	Argovia	Argovia
AI	Appenzell Innerrhoden	Appenzell Inner-Rhodes	**Appenzell Innerrhoden**	Appenzell Rhodes-Intérieures	Appenzello Interno	Appenzell dadens
AR	Appenzell Ausserrhoden	Appenzell Outer-Rhodes	**Appenzell Ausserrhoden**	Appenzell Rhodes-Extérieures	Appenzello Esterno	Appenzell dador
BS	Basel-Stadt	Basle-City	**Basel-Stadt**	Bâle-Ville	Basilea-Città	Basilea-Citad
BL	Basel-Landschaft	Basle-Country	**Basel-Landschaft**	Bâle-Campagne	Basilea-Campagna	Basilea-Champagna
BE	Bern	Berne	**Bern**	**Berne**	Berna	Berna
FR	Fribourg	Friburg	**Freiburg**	**Fribourg**	Friborgo	Friburg
GE	Geneva	-	Genf	**Genève**	Ginevra	Genevra
GL	Glarus	Glaris	**Glarus**	Glaris	Glarona	Glaruna
GR	Graubünden	Grisons	**Graubünden**	Grisons	**Grigioni**	**Grischun**
JU	Jura	-	Jura	**Jura**	Giura	Giura
LU	Lucerne	-	**Luzern**	Lucerne	Lucerna	Lucerna
NE	Neuchâtel	-	Neuenburg	**Neuchâtel**	Neuchâtel	Neuchâtel
NW	Nidwalden	Nidwald	**Nidwalden**	Nidwald	Nidvaldo	Sutsilvania
OW	Obwalden	Obwald	**Obwalden**	Obwald	Obvaldo	Sursilvania
SH	Schaffhausen	Schaffhouse	**Schaffhausen**	Schaffhouse	Sciaffusa	Schaffusa
SZ	Schwyz	-	**Schwyz**	Schwyz (or Schwytz)	Svitto	Sviz
SO	Solothurn	Soleure	**Solothurn**	Soleure	Soletta	Soloturn
SG	St. Gallen	St. Gall	**St. Gallen**	Saint-Gall	San Gallo	Son Gagl

TG	Thurgau	Thurgovia	**Thurgau**	Thurgovie	Turgovia	Turgovia
TI	Ticino	Tessin	Tessin	Tessin	**Ticino**	Tessin
UR	Uri	-	**Uri**	Uri	Uri	Uri
VS	Valais	Wallis	**Wallis**	**Valais**	Vallese	Vallais
VD	Vaud	-	Waadt	**Vaud**	Vaud	Vad
ZG	Zug	-	**Zug**	Zoug	Zugo	Zug
ZH	Zurich	-	**Zürich**	Zurich	Zurigo	Turitg

Admission of new cantons

The enlargement of Switzerland by way of the admission of new cantons ended in 1815. After a failed attempt of Vorarlberg to join Switzerland in 1919, the idea of resuming Swiss enlargement was revived in 2010 by a parliamentary motion that would allow the accession of regions bordering on Switzerland.

See also

- List of legislative and executive councils of the Cantons of Switzerland
- Data codes for Switzerland#Cantons
- List of cantons of Switzerland by elevation
- Flags of Swiss cantons
- Cantonal bank, a commercial bank (at least partially) owned by the canton

Notes

1. This is the order generally used in Swiss official documents. At the head of the list are the three city cantons that were considered preeminent in the Old Swiss Confederacy; the other cantons are listed in order of accession to the Confederation. This traditional order of precedence among the cantons has no practical relevance in the modern federal state, in which the cantons are equal to one another, although it still determines formal precedence among the cantons' officials (see Swiss order of precedence).
2. as of 5 April 2009
3. km²
4. Per km², based on 2000 population
5. As of 31 December 2007,
6. Seat of government and parliament is Herisau, the seat of the judicial authorities is Trogen
7. Seat of parliament half-yearly alternates between Frauenfeld and Weinfelden

References

- (German). Cited as *Ehrenzeller.*
- Cited as *Häfelin.*

[1] *Cantons, In the Old Confederation until 1798* in German (http://www.hls-dhs-dss.ch/textes/d/D26414-1-1.php), French (http://www.hls-dhs-dss.ch/textes/f/F26414-1-1.php) and Italian (http://www.hls-dhs-dss.ch/textes/i/I26414-1-1.php) in the online *Historical Dictionary of Switzerland.*

[2] *Jura (Canton)* in German (http://www.hls-dhs-dss.ch/textes/d/D7399.php), French (http://www.hls-dhs-dss.ch/textes/f/F7399.php) and Italian (http://www.hls-dhs-dss.ch/textes/i/I7399.php) in the online *Historical Dictionary of Switzerland.*

[3] *Cantons, In the Federal State since 1848* in German (http://www.hls-dhs-dss.ch/textes/d/D26414-1-4.php), French (http://www.hls-dhs-dss.ch/textes/f/F26414-1-4.php) and Italian (http://www.hls-dhs-dss.ch/textes/i/I26414-1-4.php) in the online *Historical Dictionary of Switzerland.*

[4] Swiss Government website (http://www.ch.ch/karte/index.html?lang=en) with links to each cantonal government, accessed 11 November 2008
[5] founding forest-canton, foundation date traditionally given as either 1307, 1304 or 1291 (see Foundation of the Old Swiss Confederacy).
[6] part of Basel until 1833/1999
[7] part of Appenzell until 1597/1999
[8] Act of Mediation, formed out of the Canton of Säntis and the northern half of the Canton of Linth.
[9] Act of Mediation; formerly the Canton of Raetia, comprising the earlier Three Leagues.
[10] coterminous with the canton of Thurgau of the Helvetic Republic (1798), formerly a condominium.
[11] combining the former cantons of Bellinzona and Lugano; see *Ennetbirgische Vogteien*.
[12] Act of Mediation, formerly Canton of Léman.
[13] Restoration, formerly the Simplon *département*
[14] claimed by Frederick William III of Prussia until the Neuchâtel Crisis of 1856–1857.
[15] seceded from Berne
[16] Federal Constitution of the Swiss Confederation of 18 April 1999, SR/RS 101 (E (http://www.admin.ch/ch/e/rs/c101.html), D (http://www.admin.ch/ch/d/sr/c101.html), F (http://www.admin.ch/ch/f/rs/c101.html), I (http://www.admin.ch/ch/i/rs/c101.html)), art. 1 (E (http://www.admin.ch/ch/e/rs/101/a1.html), D (http://www.admin.ch/ch/d/sr/101/a1.html), F (http://www.admin.ch/ch/f/rs/101/a1.html), I (http://www.admin.ch/ch/i/rs/101/a1.html))
[17] Felix Hafner / Rainer J. Schweizer in *Ehrenzeller*, Art. 1 N 2; Häfelin, N 966.
[18] Twenty-three after the creation of the Canton of Jura in 1978.
[19] Bundesverfassung der Schweizerischen Eidgenossenschaft vom 29. Mai 1874 (http://www.verfassungen.de/ch/verf74-i.htm), Bundesverfassung der Schweizerischen Eidgenossenschaft vom 12. September 1848 (http://www.verfassungen.de/ch/verf48.htm) (German); author's translation.
[20] Felix Hafner / Rainer J. Schweizer in *Ehrenzeller*, Art. 1 N 10; Häfelin, N 963
[21] Häfelin, N 963, 967
[22] Häfelin, N 950
[23] Pacte fédéral du 1er (http://www.admin.ch/org/polit/00056/index.html?lang=fr) août 1291] sur Admin.ch "vallée inférieure d'Unterwald" signifie Nidwald.
[24] Pacte fédéral du 1er août 1291 (http://hypo.ge.ch/www/cliotexte//html/suisse.histoire.1291.html) sur Cliotexte
[25] Réforme catholique, Contre-Réforme et scission (http://hls-dhs-dss.ch/textes/f/F7389-1-10.php) Article du dictionnaire historique de la Suisse
[26] De la République helvétique à la division du canton (1798-1833) (http://hls-dhs-dss.ch/textes/f/F7387-2-2.php) Article du dictionnaire historique de la Suisse

External links

- Swissworld.org (http://www.swissworld.org/en/politics/general_information/the_cantons/) – The cantons of Switzerland
- GeoPuzzle (http://www.geomit.ch/GeoPuzzles/SchweizPuzzle/schweiz.html) – Assemble cantons on a Swiss map
- Badac (http://www.badac.ch) – Database on Swiss cantons and cities (French/German)

Aargau

Aargau (German *Aargau*; rarely anglicized **Argovia**) is one of the more northerly cantons of Switzerland. It comprises the lower course of the river Aare, which is why the canton is called Aar-gau (meaning *Aare district*). It is one of the most densely populated regions of Switzerland.[]

History

Medieval Aargau

Aarau

In early medieval times, *Argovia* or *Argowe* was a disputed border region between the duchies of Alamannia and Burgundy. A line of the von Wetterau (Conradines) intermittently held the countship of Aargau from 750 until about 1030, when they lost it (having in the meantime taken the name von Tegerfelden). From the extinction in 1254 of the Hohenstaufen dynasty until 1415, the area was ruled by the Habsburgs, and many castles from that time still stand (examples include Habsburg, Lenzburg, Tegerfelden, Bobikon, Stin and Wildegg). The Habsburgs founded a number of monasteries (with some structures enduring, e.g., in Wettingen and Muri), the closing of which by the government in 1841 was a contributing factor to the outbreak of the Swiss civil war – the "Sonderbund War" – in 1847.

Under the Swiss Confederation

When Frederick IV of Habsburg sided with Antipope John XXIII at the Council of Constance, Emperor Sigismund placed him under the Imperial ban. Shortly thereafter in 1415, Bern and the rest of the Swiss Confederation used the ban as a pretext to invade Aargau. The Confederation was able to quickly conquer the towns of Aarau, Lenzburg, Brugg and Zofingen along with most of the Habsburg castles. Bern kept the southwest portion (Zofingen, Aarburg, Aarau, Lenzburg, and Brugg). Some districts, named the *Freie Ämter* (*free bailiwicks*) – Mellingen, Muri, Villmergen, and Bremgarten), with the countship of Baden – were governed as "subject lands" by all or some of the Confederates.

Unteraargau or Bernese Aargau

Bern's portion of the Aargau came to be known as the Unteraargau. In 1514 Bern expanded north into the Jura and so came into possession of several strategically important mountain passes into the Austrian Fricktal. This land was added to the Unteraargau and was directly ruled from Bern. It was divided into seven rural bailiwicks and four administrative cities, Aarau, Zofingen, Lenzburg and Brugg. While the Habsburgs were driven out, many of their minor nobles were allowed to keep their lands and offices, though over time they lost power to the Bernese government. The bailiwick administration was based on a very small staff of officials, mostly made up of Bernese citizens, but with a few locals.[1]

Districts in Bernese controlled Unteraargau

When Bern converted during the Protestant Reformation in 1528, the Unteraargau also converted. At the beginning of the 16th century a number of anabaptists migrated into the upper Wynen and Rueder valleys from Zurich. Despite pressure from the Bernese authorities in the 16th and 17th centuries anabaptism never entirely disappeared from the Unteraargau.[1]

Bern used the Aargau bailiwicks mostly as a source of grain for the rest of the city-state. The administrative cities remained economically only of regional importance. However, in the 17th and 18th centuries Bern encouraged industrial development in Unteraargau and by the late 18th century it was the most industrialized region in the city-state. The high industrialization led to high population growth in the 18tf century, for example between 1764–98, the population grew by 35%, far more than in other parts of the canton. In 1870 the proportion of farmers in Aarau, Lenzburg, Kulm, and Zofingen districts was 34-40%, while in the other districts it was 46-57%.[1]

Freie Ämter

The rest of the Freie Ämter were collectively administered as subject territories by the rest of the Confederation. Muri *Amt* was assigned to Zurich, Lucerne, Schwyz, Unterwalden, Zug and Glarus, while the *Ämter* of Meienberg, Richensee and Villmergen were first given to Lucerne alone. The final boundary was set in 1425 by an arbitration tribunal and Lucerne had to give the three *Ämter* to be collectively ruled.[2] The four *Ämter* were then consolidated under a single Confederation bailiff into what was known in the 15th century as the *Waggental* Bailiwick (German: *Vogtei im Waggental*). In the 16th century, it came to be known as the *Vogtei der Freien Ämter*. While the *Freien Ämter* often had independent lower courts, they were forced to accept the Confederation's sovereignty. Finally, in 1532 the canton of Uri became part of the collective administration of the Freien Ämter.[3]

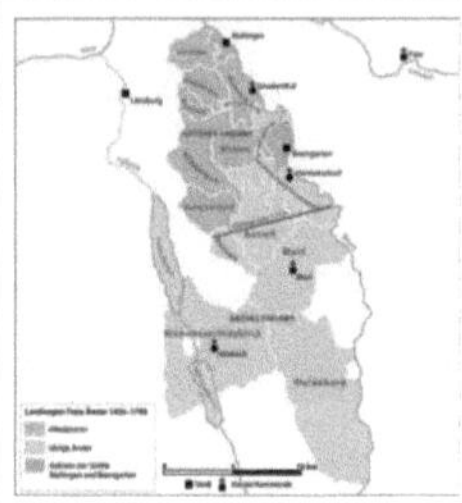

Map of the Freie Ämter, including the 1712 line dividing the Upper and Lower Freie Ämter

At the time of Reformation the majority of the Ämter converted to the new faith. In 1529 a wave of iconoclasm swept through the area and wiped away much of the old religion. After the defeat of Zurich in the second Battle of Kappel in 1531, the victorious five Catholic cantons marched their troops into the Freie Ämter and reconverted them to Catholicism.[2]

In the two Battles of Villmergen in 1656 and 1712 the Freie Ämter became the staging ground for the warring Reformed and Catholic armies. While the peace after the 1656 war did not change the status quo, the fourth Peace of Aarau in 1712 brought about a reorganization of power relations. The victory gave Zurich the opportunity to force the Catholic cantons out of the government in the county of Baden and the adjacent area of the Freie Ämter. The Freie Ämter were then divided in two by a line drawn from the gallows in Fahrwangen to the Oberlunkhofen church steeple. The northern part, the so-called Unteren Freie Ämter (lower Freie Ämter), which included the districts of Boswil (in part) and Hermetschwil and the Niederamt, were ruled by Zurich, Bern and Glarus. The southern part, the Oberen Freie Ämter (upper Freie Ämter), were ruled by the previous seven cantons but Bern was added to make an eighth.[2]

During the Helvetic Republic (1798–1803), the county of Baden, the Freie Ämter and the region known as the *Kelleramt* were combined into the Canton of Baden.

County of Baden

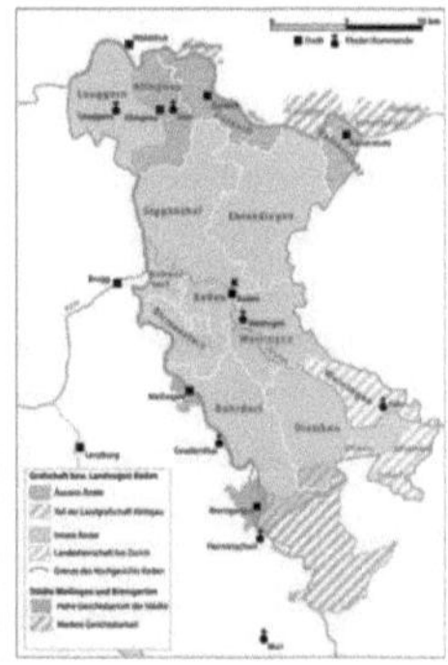

County of Baden

The County of Baden was a shared condominium of the entire Old Swiss Confederacy. After the Confederacy conquest in 1415 they retained much of the Habsburg legal structure, which caused a number of problems. The local nobility had the right to hold the low court in only about one fifth of the territory. There were over 30 different nobles who had the right to hold courts scattered around the surrounding lands. All these overlapping jurisdictions caused numerous conflicts, but gradually the Confederation was able to acquire these rights in the County. The cities of Baden, Bremgarten and Mellingen became the administrative centers and held the high courts. Together with the courts, the three administrative centers had considerable local autonomy, but were ruled by a governor who was appointed by the *Acht Orte* every two years. After the Protestant victory at the Second Battle of Villmergen, the administration of the County changed slightly. Instead of the *Acht Orte* appointing a bailiff together, Zurich and Bern each appointed the governor for 7 out of 16 years while Glarus appointed him for the remaining 2 years.[4]

The chaotic legal structure and fragmented land ownership combined with a tradition of dividing the land among all the heirs in an inheritance prevented any large scale reforms. The governor tried in the 18th century to reform and standardize laws and ownership across the County, but with limited success. With an ever changing administration, the County lacked a coherent long-term economic policy or support for reforms. By the end of the 18th century there were no factories or mills and only a few small cottage industries along the border with Zurich. Road construction first became a priority after 1750, when Zurich and Bern began appointing a governor for seven years.[4]

During the Protestant Reformation, some of the municipalities converted to the new faith. However, starting in 1531, some of the old parishes were converted back to the old faith. The governors were appointed from both Catholic and Protestant cantons and since they changed every two years, neither faith gained a majority in the County.[4]

The County was the only federal condominium in the 17th century where Jews were tolerated. In 1774, they were restricted to just two towns, Endingen and Lengnau. While the rural upper class tried several times to finally expel the Jews, the financial interests of the authorities prevented this. The Jews were directly subordinate to the governor starting in 1696 when they were forced to buy a protecting and shielding letter every 16 years from the governor.[4]

After the French invasion, on 19 March 1798, the governments of Zurich and Bern agreed to the creation of the short lived Canton of Baden in the Helvetic Republic. With the Act of Mediation in 1803, the Canton of Baden was dissolved. Portions of the lands of the former County of Baden now became the District of Baden in the newly created Canton of Aargau. After World War II, this formerly agrarian region saw striking growth and became the district with the largest and densest population in the Canton (110,000 in 1990, 715 persons per km2).[4]

The canton of Aargau

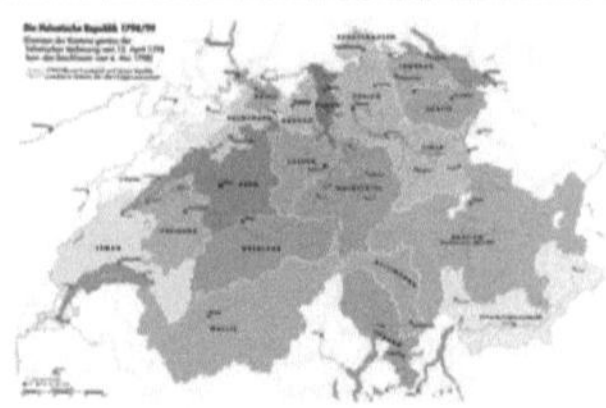
Helvetic Republic on 1798/99, with the Cantons of Aargau and Baden

French forces occupied Aargau from 10 March to 18 April 1798; thereafter the Bernese portion became the canton of Aargau in the Helvetic Republic and the remainder formed the Canton of Baden. In 1803, the two halves were united under the name of canton of Aargau,[] [5] which was then admitted as a full member of the reconstituted Confederation. Fricktal, ceded in 1802 by Austria via Napoleonic France to the Helvetic Republic, was briefly a separate Swiss canton under a *Statthalter* ('Lieutenant'), but on 9 March 1803 was incorporated in the canton of Aargau.

The chief magistracy of Aargau changed its style repeatedly:

- first two consecutive *Regierungsstatthalter* :
  - April 1798 – November 1801 Jakob Emmanuel Feer (b. 1754 – d. 1833)
  - 1802–1803 Johann Heinrich Rothpletz (b. 1766 – d. 1833)
- Presidents of the Government Commission
  - 10 March 1803 – 26 April 1803 Johann Rudolf Dolder (b. 1753 – d. 1807)
  - 26 April 1803 – 1815 a 'Small Council' (president rotating monthly)
- annual *Amtsbürgermeister* 1815–1831
- annual *Landammänner* since 1815

In the year 2003, the canton of Aargau celebrated its 200th anniversary.

The Jewish history in Aargau

Two separate doors (one for Jews and one for Christians) on a house in Lengnau

In the 17th century, Jews were banished from Switzerland. However, a few families were permitted to live in two villages, Endingen and Lengnau, in Aargau which became the Jewish ghetto in Switzerland. This remained the case until the 19th century. In 1799 all special tolls were abolished, and in 1802 the poll tax was removed. On 5 May 1809, the right of citizenship was granted to Jews, and they were permitted to engage in trade and agriculture. The right of settlement remained restricted to Endingen and Lengnau until 7 May 1846, when they were allowed to settle in any portion of the canton of Aargau. Ten years later (24 September 1856) the Swiss Federal Council voted them equal political rights with other Swiss citizens in that canton, as well as entire freedom of commerce; but the opposition of the Christian population prevented the decision from being generally carried out. The federal authorities in July 1863, granted the Swiss Jews the fullest rights of citizens. However, full civil equality was obtained only when they received the formal rights of citizenship, which had long been withheld from them in their own communities of Endingen and Lengnau. A resolution of the Grand Council, on 15 May 1877, granted citizens' rights to the members of the Jewish communities of those places, giving them charters under the names of New Endingen and New Lengnau.[6]

Geography

View of the Lägern from Bözberg

Its total area is 1404 square kilometers (sq mi). The capital is Aarau. It borders Germany to the north. To the west lie the cantons of Basel-Landschaft, Solothurn and Bern. The canton of Lucerne lies south of the canton of Aargau, Zürich and Zug to the east.

The canton of Aargau is one of the least mountainous Swiss cantons, forming part of a great table-land, to the north of the Alps and the east of the Jura, above which rise low hills. The surface of the country is beautifully diversified, undulating tracts and well-wooded hills[7] alternating with fertile valleys watered mainly by the Aar and its tributaries. The valleys alternate with pleasant hills, most of which are full of woods. Slightly over one-third of the canton is wooded (518 square kilometers (sq mi)), while nearly half is used from farming (635.7 square kilometers (sq mi)). 33.5 square kilometers (sq mi) or about 2.4% of the canton is considered unproductive, mostly lakes and streams. The canton is heavily developed, with 216.7 square kilometers (sq mi) or about 15% of the canton developed for housing or transportation.[8]

It contains the famous hot sulphur springs of Baden and Schinznach-Bad, while at Rheinfelden there are very extensive saline springs. Just below Brugg the Reuss and the Limmat join the Aar, while around Brugg are the ruined castle of Habsburg, the old convent of Königsfelden (with fine painted medieval glass) and the remains of the Roman settlement of *Vindonissa* (Windisch).

Political subdivisions

Districts

Districts in Aargau

Aargau is divided into 11 districts:

- Aarau with capital Aarau
- Baden with capital Baden
- Bremgarten with capital Bremgarten
- Brugg with capital Brugg
- Kulm with capital Unterkulm
- Laufenburg with capital Laufenburg
- Lenzburg with capital Lenzburg
- Muri with capital Muri
- Rheinfelden with capital Rheinfelden
- Zofingen with capital Zofingen

- Zurzach with capital Zurzach

Municipalities

There are a total of 231 municipalities in the canton of Aargau.

Coat of arms

The blazon of the coat of arms is *Per pale, dexter: sable, a fess wavy argent, charged with two cotises wavy azure; sinister: sky blue, three mullets of five argent.*[9]

Demographics

Aargau has a population (as of December 2012) of 627893.[10] As of 2010, 21.5% of the population are resident foreign nationals. Over the last 10 years (2000–2010) the population has changed at a rate of 11%. Migration accounted for 8.7%, while births and deaths accounted for 2.8%.[11] Most of the population (as of 2000) speaks German (477,093 or 87.1%) as their first language, Italian is the second most common (17,847 or 3.3%) and Serbo-Croatian is the third (10,645 or 1.9%). There are 4,151 people who speak French and 618 people who speak Romansh.[12]

Of the population in the canton, 146,421 or about 26.7% were born in Aargau and lived there in 2000. There were 140,768 or 25.7% who were born in the same canton, while 136,865 or 25.0% were born somewhere else in Switzerland, and 107,396 or 19.6% were born outside of Switzerland.[12]

As of 2000, children and teenagers (0–19 years old) make up 24.3% of the population, while adults (20–64 years old) make up 62.3% and seniors (over 64 years old) make up 13.4%.[11]

As of 2000, there were 227,656 people who were single and never married in the canton. There were 264,939 married individuals, 27,603 widows or widowers and 27,295 individuals who are divorced.[12]

As of 2000, there were 224,128 private households in the canton, and an average of 2.4 persons per household.[11] There were 69,062 households that consist of only one person and 16,254 households with five or more people. As of 2009, the construction rate of new housing units was 6.5 new units per 1000 residents.[11] The vacancy rate for the canton, in 2010, was 1.54%.[11]

The majority of the population is centered on one of three areas: the Aare Valley, the side branches of the Aare Valley, or along the Rhine River.[]

Historic population

The historical population is given in the following chart:[13] [14] [15]

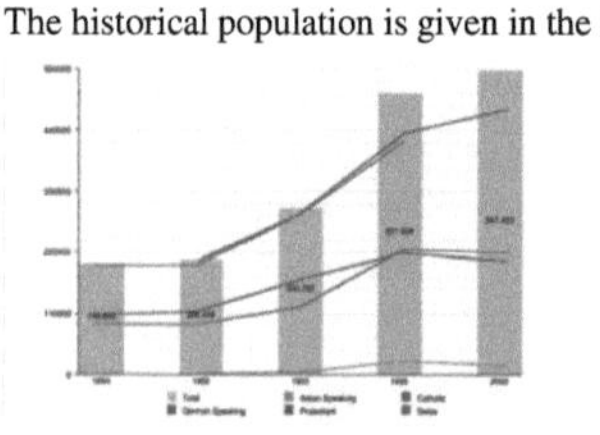

Politics

In the 2011 federal election the most popular party was the SVP which received 34.7% of the vote. The next three most popular parties were the SP/PS (18.0%), the FDP (11.5%) and the CVP (10.6%).[16]

The SVP received about the same percentage of the vote as they did in the 2007 Federal election (36.2% in 2007 vs 34.7% in 2011). The SPS retained about the same popularity (17.9% in 2007), the FDP retained about the same popularity (13.6% in 2007) and the CVP retained about the same popularity (13.5% in 2007).[17]

Religion

Catholic City Church in Baden, Aargau

From the 2000 census, 219,800 or 40.1% were Roman Catholic, while 189,606 or 34.6% belonged to the Swiss Reformed Church. Of the rest of the population, there were 11,523 members of an Orthodox church (or about 2.10% of the population), there were 3,418 individuals (or about 0.62% of the population) who belonged to the Christian Catholic Church, and there were 29,580 individuals (or about 5.40% of the population) who belonged to another Christian church. There were 342 individuals (or about 0.06% of the population) who were Jewish, and 30,072 (or about 5.49% of the population) who were Islamic. There were 1,463 individuals who were Buddhist, 2,089 individuals who were Hindu and 495 individuals who belonged to another church. 57,573 (or about 10.52% of the population) belonged to no church, are agnostic or atheist, and 15,875 individuals (or about 2.90% of the population) did not answer the question.[12]

Education

In Aargau about 212,069 or (38.7%) of the population have completed non-mandatory upper secondary education, and 70,896 or (12.9%) have completed additional higher education (either university or a *Fachhochschule*). Of the 70,896 who completed tertiary schooling, 63.6% were Swiss men, 20.9% were Swiss women, 10.4% were non-Swiss men and 5.2% were non-Swiss women.[12]

Economy

Leibstadt Nuclear Power Plant

As of 2010, Aargau had an unemployment rate of 3.6%. As of 2008, there were 11,436 people employed in the primary economic sector and about 3,927 businesses involved in this sector. 95,844 people were employed in the secondary sector and there were 6,055 businesses in this sector. 177,782 people were employed in the tertiary sector, with 21,530 businesses in this sector.[11]

In 2008 the total number of full-time equivalent jobs was 238,225. The number of jobs in the primary sector was 7,167, of which 6,731 were in agriculture, 418 were in forestry or lumber production and 18 were in fishing or fisheries. The number of jobs in the secondary sector was 90,274 of which 64,089 or (71.0%) were in manufacturing, 366 or (0.4%) were in mining and 21,705 (24.0%) were in construction. The number of jobs in the tertiary sector was 140,784. In the tertiary sector; 38,793 or 27.6% were in the sale or repair of motor vehicles, 13,624 or 9.7% were in the movement and storage of goods, 8,150 or 5.8% were in a hotel or restaurant, 5,164 or 3.7% were in the information industry, 5,946 or 4.2% were the insurance or financial industry, 14,831 or 10.5% were technical professionals or scientists, 10,951 or 7.8% were in education and

21,952 or 15.6% were in health care.[18]

Of the working population, 19.5% used public transportation to get to work, and 55.3% used a private car.[11] Public transportation – bus and train – is provided by Busbetrieb Aarau AG.

The farmland of the canton of Aargau is some of the most fertile in Switzerland. Dairy farming, cereal and fruit farming are among the canton's main economic activities. The canton is also industrially developed, particularly in the fields of electrical engineering, precision instruments, iron, steel and cement.

Three of Switzerland's five nuclear power plants are in the canton of Aargau (Beznau I + II and Leibstadt). Additionally, the many rivers supply enough water for numerous hydroelectric power plants throughout the canton. The canton of Aargau is often called "the energy canton".

A significant number of people commute into the financial center of the city of Zürich, which is just across the cantonal border. As such the per capita cantonal income (in 2005) is 49,209 CHF.[19]

Tourism is significant, particularly for the hot springs at Baden and Schinznach-Bad, the ancient castles, the landscape, and the many old museums in the canton.[7] Hillwalking is another tourist attraction but is of only limited significance.

See also

- Canton of Aargau castles and fortresses
- Aargauer Zeitung
- FC Aarau
- GP Kanton Aargau, bicycle race

References

[1] *Berner Aargau* in German (http://www.hls-dhs-dss.ch/textes/d/D29288.php), French (http://www.hls-dhs-dss.ch/textes/f/F29288.php) and Italian (http://www.hls-dhs-dss.ch/textes/i/I29288.php) in the online *Historical Dictionary of Switzerland.*

[2] *Freie Ämter* in German (http://www.hls-dhs-dss.ch/textes/d/D8298.php), French (http://www.hls-dhs-dss.ch/textes/f/F8298.php) and Italian (http://www.hls-dhs-dss.ch/textes/i/I8298.php) in the online *Historical Dictionary of Switzerland.*

[3] Adolf Gasser: *Die territoriale Entwicklung der Schweizerischen Eidgenossenschaft 1291–1797.* Sauerländer: Aarau 1932, p. 82 **(German)**

[4] *Baden (AG), County* in German (http://www.hls-dhs-dss.ch/textes/d/D8296.php), French (http://www.hls-dhs-dss.ch/textes/f/F8296.php) and Italian (http://www.hls-dhs-dss.ch/textes/i/I8296.php) in the online *Historical Dictionary of Switzerland.*

[5] Bridgwater, W. & Beatrice Aldrich. (1966) *The Columbia-Viking Desk Encyclopedia.* Columbia University. p. 11.

[6] Jewish Encyclopedia: Aargau (http://www.jewishencyclopedia.com/view.jsp?artid=3&letter=A)

[7] "Aargau" (http://search.eb.com/eb/article-9003207) *Encyclopædia Britannica.* 11 January 2007

[9] Flags of the World.com (http://www.crwflags.com/fotw/flags/ch-ag.html) accessed 26-January-2012

[10] Canton of Aargau Statistical Office (https://www.ag.ch/de/dfr/statistik/statistische_daten/oeffentliche_statistik/bevoelkerung/bevoelkerung_1.jsp), MS Excel document – *Bevölkerungsbestand 2012 nach Gemeinde, Geschlecht u. Nationalität (per 31. Dezember 2012)*] **(German)** accessed 23 July 2013

[11] Swiss Federal Statistical Office (http://www.bfs.admin.ch/bfs/portal/en/index/regionen/regionalportraets.html) accessed 26-January-2012

[12] STAT-TAB Datenwürfel für Thema 40.3 - 2000 (http://www.pxweb.bfs.admin.ch/Database/German_40 - EidgenÃ¶ssische VolkszÃ¤hlung/40.3 - 2000/40.3 - 2000.asp?lang=1&prod=40&secprod=3&openChild=true) **(German)** accessed 2 February 2011

[13] *Aargau* in German (http://www.hls-dhs-dss.ch/textes/d/D7392.php), French (http://www.hls-dhs-dss.ch/textes/f/F7392.php) and Italian (http://www.hls-dhs-dss.ch/textes/i/I7392.php) in the online *Historical Dictionary of Switzerland.*

[14] Swiss Federal Statistical Office - Religion (http://www.bfs.admin.ch/bfs/portal/de/index/themen/01/05/blank/key/religionen.html) accessed 26 January 2012

[15] Swiss Federal Statistical Office - Languages (http://www.bfs.admin.ch/bfs/portal/de/index/themen/01/05/blank/key/sprachen.html) accessed 26 January 2012

[16] Swiss Federal Statistical Office, Elections in Switzerland (http://www.portal-stat.admin.ch/nrw/files/de/01.xml) **(German)** accessed 5 January 2012

[17] Swiss Federal Statistical Office, *Nationalratswahlen 2007: Stärke der Parteien und Wahlbeteiligung, nach Gemeinden/Bezirk/Canton* (http://www.bfs.admin.ch/bfs/portal/de/index/themen/17/02/blank/data/04/03.html) **(German)** accessed 28 May 2010

[18] Swiss Federal Statistical Office STAT-TAB Betriebszählung: Arbeitsstätten nach Gemeinde und NOGA 2008 (Abschnitte), Sektoren 1-3 (http://www.pxweb.bfs.admin.ch/Database/German_06 - Industrie und Dienstleistungen/06.2 - Unternehmen/06.2 - Unternehmen.asp?lang=1&prod=06&secprod=2&openChild=true) (German) accessed 28 January 2011
[19] Swiss Federal Statistical Office - Per capita cantonal income (http://www.bfs.admin.ch/bfs/portal/en/index/themen/04/02/05/key/pro-kopf-einkommen.html) accessed 26 January 2012

External links

- Official website (http://www.ag.ch) (German)
- *Aargau* in German (http://www.hls-dhs-dss.ch/textes/d/D7392.php), French (http://www.hls dhs-dss.ch/textes/f/F7392.php) and Italian (http://www.hls-dhs-dss.ch/textes/i/I7392.php) in the online *Historical Dictionary of Switzerland.*
- Swiss Federal Statistical Office-Aargau Statistics (http://www.bfs.admin.ch/bfs/portal/en/index/regionen/regionalportraets/aargau/blank/kennzahlen.html)

Basel-Landschaft

Basel-Landschaft (rarely **Basle-Campagne**; German: *Basel-Landschaft* "Basel-Country"; informally: *Baselland/Baselbiet*), is one of the 26 cantons of Switzerland. The capital is Liestal. It shares borders with the cantons of Basel-Stadt, Solothurn, Jura and Aargau, and with the French région of Alsace and the German state of Baden-Württemberg.

History

Basel-Landschaft, together with Basel-Stadt, formed the historic Canton of Basel until they separated following the uprising of 1833 (Battle of the Hülftenschanz near Frenkendorf).

Castle of Angenstein in Duggingen

In Roman times, the area of Basel was a centre of Roman activity. There are well-preserved remains at the site of *Augusta Raurica* in the canton of Basel-Landschaft. Around AD 200 there were about 20,000 people living in this city, now part of the much smaller Augst. The remains are on display in an open-air museum. The museum attracts over 140,000 visitors per year. Many of these visitors are schoolchildren from other parts of Switzerland. The site of Augusta Raurica includes the best-preserved amphitheatre north of the Alps, and a reconstructed Roman villa.

The lands of the canton Basel-Landschaft are part of the lands acquired by the city of Basel. Until the end of the 16th century, most of the canton's land belonged to the city of Basel. After Napoleon's visit in 1798, the country achieved equality with the city. The country was economically dependent on the city, most probably because of the low level of education in the agricultural areas at the time. The city of Basel remained the cultural and economic centre of both Basel half cantons until then. Castles and residences of Basel merchants dominated much of the landscape in Basel-Country.

Caricature of the division of Basel, 1833

After 1830 there were political quarrels and armed conflict in the canton of Basel. Some of these were concerned with the rights of the population in the agricultural areas. They ultimately led to the separation of the canton Basel-Landschaft from the city of Basel on 26 August 1833. Since then, there has been a movement for reunification. This movement gained momentum after 1900 when many parts of Basel-Landschaft became industrialized. The two half cantons agreed in principle to merge, but in 1969 the people of Basel-Landschaft voted down a referendum on this proposal in favour of retaining their independence. It is thought that the closing economic gap between the two cantons was the main reason why the population changed their attitude.

That vote was not the end of a close relationship between the two Basels. The two half cantons have since signed a number of agreements to co-operate. The contribution of Basel-Landschaft to the University of Basel since 1976 is just one example.

Geography

The canton of Basel-Landschaft lies in the north of Switzerland. The Jura mountain chain traverses the canton. The rivers Ergolz and Birs drain the lands of the canton. The thirty rivers of Baselland are :

Ergolz, Eibach, Homburger Bach, Diegter Bach, Walibach, Vodere Frenke, Fluebach, Hintere Frenke, Frenke, Orisbach, Röserenbach, Arisdorfer Bach, Violenbach, Wintersinger Bach, Buuser Bach, Magdener Bach, Hemmiker Bach, Rickenbächlein, Rhein, Birsig, Binnbach, Marchbach, Birs, Seebach, Chastelbach, Ibach, Dürrbach, Lüssel, Wahlenbach and Lützel.

Politics

Former town hall in Liestal, the cantonal capital.

Since the decision to remain independent from Basel-Stadt in 1969 there have been a number of requests in the parliament for Basel-Landschaft to become a *full* canton. In 1988 the canton of Basel-Landschaft had this aim written into its constitution. The aim remains to change the Swiss constitution to recognize the two cantons of Basel as full members.

Political subdivisions

The nine municipalities of the *Arlesheim* district used to belong to the diocese of Basel. In 1792 French troops occupied the district and in 1793 the lands were annexed by France, which explains the linguistic switch of the Family of the House of Basel's name from "von Basel" to "de Bâle", since Arlesheim holds their manor. In 1815 at the Congress of Vienna the district joined Basel.

The district of *Laufental* has the same history as that of *Arlesheim.* The important difference is that in 1815 Laufental joined Bern rather than Basel. When the canton of Jura was created in 1979, the district of Laufental became an enclave of the canton of Bern. It was allowed self- determination and in 1980 the people decided to join the canton of Basel-Landschaft. This led to *Laufental* joining the canton of Basel-Landschaft on 1 January 1994 after a lengthy administrative process.

Districts

There are 5 areas (Bezirke) in Basel-Landschaft (as of 2012):

- Arlesheim (*Pop:* 151156) capital: Arlesheim (*Pop:* 9096)
- Laufen (*Pop:* 19162) capital: Laufen (*Pop:* 5351)
- Liestal (*Pop:* 57833) capital: Liestal (*Pop:* 13771)
- Sissach (*Pop:* 34176) capital: Sissach (*Pop:* 6296)
- Waldenburg (*Pop:* 15646) capital: Waldenburg (*Pop:* 1196)

districts of Canton Basel-Landschaft

Municipalities

There are 86 municipalities in the canton (As of 2009).[]

Demographics

The population is predominantly German speaking. Protestantism is the main religion in the canton (43% as of 2000), while about one-third of the population (32%) is Roman Catholic.[1] The Swiss Reformed Church and Christian Catholic Church of Switzerland are recognized as state churches. Between 1959 and 1970 the canton was Switzerland's fastest-growing, as the population almost doubled from 108,000 to 205,000. Today, the canton has a population (as of 31 December 2012) of 277973.[2] As of 2007, the population included 48,719 foreigners who made up 18.1% of the population.[3]

Economy

Agriculture in the canton includes fruit growing, dairy farming and cattle breeding. Important industries include textiles, metals and chemicals.

Elektra Birseck Münchenstein in Münchenstein

The canton of Basel-Landschaft is part of the economic region around Basel that includes parts of France and Germany as well as both half cantons of Basel. Since the 1960s there are agreements in force to strengthen contacts within the so-called *Regio Basiliensis*. This economic co-operation is often considered as the most intensive in Europe.

From the 17th century until the beginning of the 20th century silk weaving was important in Basel-Landschaft. Factories were established as early as 1850, following the finding of salt in underground deposits, founding industries such as the chemical industry in *Schweizerhalle*. The chemical industry means that Basel is one of the richer parts of Switzerland.

The chemical industry is not the sole employer in the canton. Small and middle-sized businesses make up about a quarter of the cantonal economy. Many of these, however, are connected to the larger employers.

Notes and references

[2] Canton of Basel-Land Statistics (http://www.statistik.bl.ch/stabl_data/stabl_bv/index.php?thema_id=2&unterthema_id=58), *Wohnbevölkerung nach Nationalität und Konfession per 31. Dezember 2012* (German) accessed 26 June 2013

See also

- Augusta Raurica
- List of castles and fortresses in Switzerland
- Basellandschaftliche Zeitung, daily newspaper published in Liestal.

External links

- Official Page (http://www.baselland.ch) (German)
- Official Statistics (http://www.statistik.admin.ch/stat_ch/ber00/ekan_bl.htm)
- *Basel-Landschaft* in German (http://www.hls-dhs-dss.ch/textes/d/D7477.php), French (http://www.hls-dhs-dss.ch/textes/f/F7477.php) and Italian (http://www.hls-dhs-dss.ch/textes/i/I7477.php) in the online *Historical Dictionary of Switzerland.*

Fricktal

The **Fricktal** ("Frick Valley") region is the northwest finger extending from the Swiss canton of Aargau. It is situated in Northwestern Switzerland east of Basel, between the High Rhine forming the border with Germany in the north and the Jura Mountains in the south. The Fricktal contains the districts of Laufenburg and Rheinfelden. The region is named after its former chief village Frick.

Fricktal coat of arms

In the Early Middle Ages, Fricktal was part of the Alamannian *Augstgau* shire between the Rhine and Aar rivers, from the 10th century onwards of the smaller *Frickgau* region within Upper Burgundy. The western Fricktal was held by the Burgundian Lords of Rheinfelden, their last scion Rudolf of Rheinfelden became Duke of Swabia in 1057 and upon his death in 1080 his possessions passed to his son in law Berthold II from the House of Zähringen. After the extinction of the line in 1218, the area eventually passed to the Counts of Habsburg, who already held the *Vogtei* of Laufenburg.

After the Habsburg dynasty had lost large parts of its original Swabian possessions south of the Rhine to the Swiss Confederacy at the 1386 Battle of Sempach, the remaining Fricktal was administered from the *Oberamt* Breisgau of Further Austria (*Vorderösterreich*) at Freiburg, while the adjacent Unteraargau region to the south was finally conquered by the Swiss at Bern in 1415. In 1469 indebted Archduke Sigismund of Further Austria sold the Breisgau with Fricktal to the Burgundian Duke Charles the Bold, neverteheless upon Charles' death in 1477 it fell back to Archduke Maximilian I of Habsburg by marriage with the duke's daughter Mary the Rich.

In the early 18th century, the Bern citizens attempted to purchase Fricktal from the Habsburg Emperors without success. Subsequent to the 1797 Treaty of Campo Formio, Fricktal was occupied by France and in 1802 it was briefly a canton of the Helvetic Republic, as the Canton of Fricktal. On 9 March 1803, Fricktal was integrated into the Aargau canton of the Swiss Confederation by order of Napoleon.

Today Fricktal with its mild climate is an important location for wine and cherry production as well as a base of chemical and pharmaceutical industries.

Obermumpf

Obermumpf	
Country	Switzerland
Canton	Aargau
District	Rheinfelden

Population	998 (Dec 2012)[1]
- Density	198 /km² (514 /sq mi)
Area	km² (sq mi)[2]
Elevation	377 m (ft)
Postal code	4234
SFOS number	4256
Surrounded by	Hellikon, Mumpf, Münchwilen, Schupfart, Stein, Zuzgen
Website	www.obermumpf.ch [3] SFSO statistics [4]

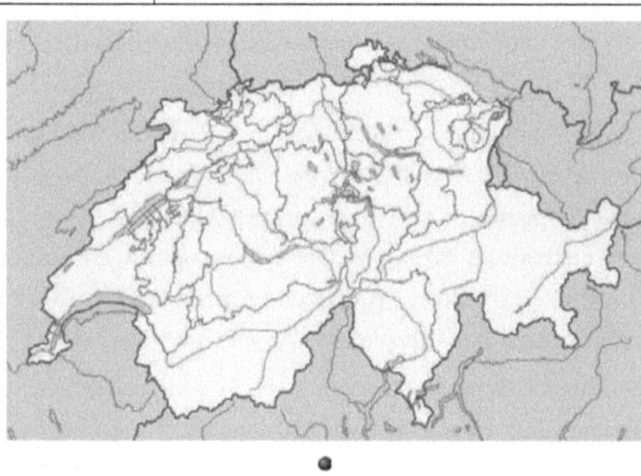

Obermumpf

Obermumpf is a municipality in the district of Rheinfelden in the canton of Aargau in Switzerland.

History

It is likely that there was a Hallstatt culture settlement near Obermumpf, though the exact sight has not been discovered. Near the village church the ruins of a Roman era farm were discovered. On the outskirts of the village, several Alamanni tombs were also found. The modern municipality of Obermumpf is first mentioned around 1302-04 as *Obermumphier*.[3] Until 1797 it was part of the Austrian Habsburgs district of Rheinfelden. After the Act of Mediation in 1803, the entire Fricktal, including Obermumpf, went to the newly formed Canton of Aargau. The low justice right was held by the manor house *Stein*, which was owned by Säckingen Abbey. Later those rights transferred to the Austrians.[3]

The Church of St. Peter and Paul, whose patronage rights originally belonged to the collegiate church of Säckingen, has a tower from 1494 and a nave from 1738. The original church, from the 8th Century, was built on the foundations of a Roman estate. Since 1889 it has been the center of the Christian Catholic parish of Obermumpf-Wallbach. In 1893-94 a catholic church was constructed, and was rebuilt in 1962.[3]

Geography

Obermumpf has an area, as of 2009, of 5.03 square kilometers (sq mi). Of this area, 2.65 km^2 (sq mi) or 52.7% is used for agricultural purposes, while 1.92 km^2 (sq mi) or 38.2% is forested. Of the rest of the land, 0.46 km^2 (sq mi) or 9.1% is settled (buildings or roads), 0.01 km^2 (acres) or 0.2% is either rivers or lakes.[4]

Of the built up area, housing and buildings made up 6.0% and transportation infrastructure made up 2.8%. Out of the forested land, 36.6% of the total land area is heavily forested and 1.6% is covered with orchards or small clusters of trees. Of the agricultural land, 30.4% is used for growing crops and 17.9% is pastures, while 4.4% is used for orchards or vine crops. All the water in the municipality is in rivers and streams.[4]

The municipality is located in the Rheinfelden district, in the Fricktal (Frick river valley). It consists of the linear village of Obermumpf.

Coat of arms

The blazon of the municipal coat of arms is *Or a Key and a Sword in saltire both Azure handled Sable.*[5]

Demographics

Obermumpf has a population (as of December 2012) of 998[1] As of June 2009, 10.6% of the population are foreign nationals.[6] Over the last 10 years (1997–2007) the population has changed at a rate of 1.3%. Most of the population (as of 2000) speaks German(93.0%), with Albanian being second most common (3.5%) and French being third (1.3%).[7]

The age distribution, as of 2008, in Obermumpf is; 95 children or 9.4% of the population are between 0 and 9 years old and 147 teenagers or 14.5% are between 10 and 19. Of the adult population, 99 people or 9.8% of the population are between 20 and 29 years old. 125 people or 12.3% are between 30 and 39, 204 people or 20.1% are between 40 and 49, and 142 people or 14.0% are between 50 and 59. The senior population distribution is 98 people or 9.7% of the population are between 60 and 69 years old, 67 people or 6.6% are between 70 and 79, there are 33 people or 3.3% who are between 80 and 89,and there are 4 people or 0.4% who are 90 and older.[8]

As of 2000 the average number of residents per living room was 0.59 which is about equal to the cantonal average of 0.57 per room. In this case, a room is defined as space of a housing unit of at least 4 m^2 (sq ft) as normal bedrooms, dining rooms, living rooms, kitchens and habitable cellars and attics.[9] About 61% of the total households were owner occupied, or in other words did not pay rent (though they may have a mortgage or a rent-to-own agreement).[10]

As of 2000, there were 22 homes with 1 or 2 persons in the household, 194 homes with 3 or 4 persons in the household, and 153 homes with 5 or more persons in the household.[11] As of 2000, there were 384 private households (homes and apartments) in the municipality, and an average of 2.5 persons per household.[7] In 2008 there were 186 single family homes (or 42.8% of the total) out of a total of 435 homes and apartments.[12] There were a total of 11 empty apartments for a 2.5% vacancy rate.[12] As of 2007, the construction rate of new housing units was 2.9 new units per 1000 residents.[7]

In the 2007 federal election the most popular party was the SVP which received 32.79% of the vote. The next three most popular parties were the SP (21.59%), the CVP (21.17%) and the FDP (7.93%). In the federal election, a total of 311 votes were cast, and the voter turnout was 42.7%.[13]

The historical population is given in the following table:[3] [8]

Historical population		
Year	**Pop.**	**±%**
1768	267	—
1850	508	+90.3%
1900	503	−1.0%
1950	595	+18.3%
1990	850	+42.9%
2000	980	+15.3%

Economy

As of 2007, Obermumpf had an unemployment rate of 1.68%. As of 2005, there were 42 people employed in the primary economic sector and about 16 businesses involved in this sector. 43 people are employed in the secondary sector and there are 13 businesses in this sector. 78 people are employed in the tertiary sector, with 18 businesses in this sector.[7]

In 2000 there were 508 workers who lived in the municipality. Of these, 404 or about 79.5% of the residents worked outside Obermumpf while 46 people commuted into the municipality for work. There were a total of 150 jobs (of at least 6 hours per week) in the municipality.[14] Of the working population, 12% used public transportation to get to work, and 61.6% used a private car.[7]

Religion

From the 2000 census, 551 or 56.2% were Roman Catholic, while 160 or 16.3% belonged to the Swiss Reformed Church. Of the rest of the population, there were 138 individuals (or about 14.08% of the population) who belonged to the Christian Catholic faith.[11]

Education

In Obermumpf about 77.1% of the population (between age 25-64) have completed either non-mandatory upper secondary education or additional higher education (either university or a *Fachhochschule*).[7] Of the school age population (in the 2008/2009 school year), there are 83 students attending primary school in the municipality.[11]

References

[1] Canton of Aargau Statistical Office (https://www.ag.ch/de/dfr/statistik/statistische_daten/oeffentliche_statistik/bevoelkerung/bevoelkerung_1.jsp), MS Excel document – *Bevölkerungsbestand 2012 nach Gemeinde, Geschlecht u. Nationalität (per 31. Dezember 2012)*] (German) accessed 23 July 2013

[2] Arealstatistik Standard - Gemeindedaten nach 4 Hauptbereichen (http://www.bfs.admin.ch/bfs/portal/de/index/themen/02/03/blank/data/gemeindedaten.html)

[3] *Obermumpf* in German (http://www.hls-dhs-dss.ch/textes/d/D1801.php), French (http://www.hls-dhs-dss.ch/textes/f/F1801.php) and Italian (http://www.hls-dhs-dss.ch/textes/i/I1801.php) in the online *Historical Dictionary of Switzerland.*

[4] Swiss Federal Statistical Office-Land Use Statistics (http://www.bfs.admin.ch/bfs/portal/de/index/themen/02/03/blank/data/gemeindedaten.html) 2009 data (German) accessed 25 March 2010

[5] Flags of the World.com (http://www.fotw.us/flags/ch-ag182.html) accessed 28-May-2010

[6] Statistical Department of Canton Aargau -Bereich 01 -Bevölkerung (http://www.ag.ch/staag/) (German) accessed 20 January 2010

[7] Swiss Federal Statistical Office (http://www.bfs.admin.ch/bfs/portal/en/index/regionen/02/key.html) accessed 28-May-2010

[8] Statistical Department of Canton Aargau -Bevölkerungsdaten für den Kanton Aargau und die Gemeinden (Archiv) (http://www.ag.ch/staag/) (German) accessed 20 January 2010

[10] Urban Audit Glossary pg 17

[11] Statistical Department of Canton Aargau - Aargauer Zahlen 2009 (http://www.ag.ch/staag/) (German) accessed 20 January 2010

[12] Statistical Department of Canton Aargau (http://www.ag.ch/staag/) (German) accessed 20 January 2010

[13] Swiss Federal Statistical Office, *Nationalratswahlen 2007: Stärke der Parteien und Wahlbeteiligung, nach Gemeinden/Bezirk/Canton* (http://www.bfs.admin.ch/bfs/portal/de/index/themen/17/02/blank/data/04/03.html) (German) accessed 28 May 2010

[14] Statistical Department of Canton Aargau-Bereich 11 Verkehr und Nachrichtenwesen (http://www.ag.ch/staag/) (German) accessed 21 January 2010

Schupfart

Schupfart	

Country Switzerland Canton Aargau District Rheinfelden	
Population	759 (Dec 2012)[1]
- Density	108 /km^2 (279 /sq mi)
Area	km^2 (sq mi)[2]
Elevation	457 m (ft)
Postal code	4325
SFOS number	4259
Surrounded by	Eiken, Frick, Gipf-Oberfrick, Hellikon, Münchwilen, Obermumpf, Wegenstetten
Website	www.schupfart.ch [3] SFSO statistics [4]

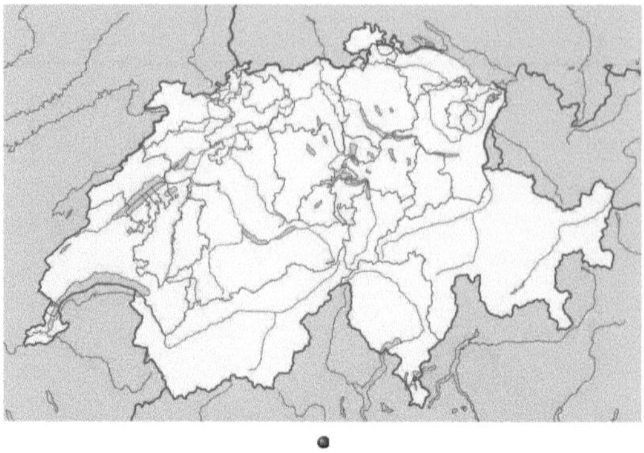

Schupfart

Schupfart is a municipality in the district of Rheinfelden in the canton of Aargau in Switzerland.

Geography

Schupfart has an area, as of 2009, of 7.04 square kilometers (sq mi). Of this area, 4.23 km^2 (sq mi) or 60.1% is used for agricultural purposes, while 2.28 km^2 (sq mi) or 32.4% is forested. Of the rest of the land, 0.49 km^2 (sq mi) or 7.0% is settled (buildings or roads), 0.01 km^2 (acres) or 0.1% is either rivers or lakes.[3]

Of the built up area, housing and buildings made up 2.7% and transportation infrastructure made up 3.6%. Out of the forested land, 30.5% of the total land area is heavily forested and 1.8% is covered with orchards or small clusters of trees. Of the agricultural land, 39.1% is used for growing crops and 16.5% is pastures, while 4.5% is used for orchards or vine crops. All the water in the municipality is in rivers and streams.[3]

Coat of arms

The blazon of the municipal coat of arms is *Argent a Linden-leaf slipped Vert.*[4]

Demographics

Schupfart has a population (as of December 2012) of 759[1] As of June 2009, 8.2% of the population are foreign nationals.[5] Over the last 10 years (1997–2007) the population has changed at a rate of 13.7%. Most of the population (as of 2000) speaks German(96.5%), with Albanian being second most common (2.3%) and Italian being third (0.6%).[6]

The age distribution, as of 2008, in Schupfart is; 78 children or 10.2% of the population are between 0 and 9 years old and 91 teenagers or 12.0% are between 10 and 19. Of the adult population, 96 people or 12.6% of the population are between 20 and 29 years old. 103 people or 13.5% are between 30 and 39, 145 people or 19.1% are between 40 and 49, and 108 people or 14.2% are between 50 and 59. The senior population distribution is 70 people or 9.2% of the population are between 60 and 69 years old, 36 people or 4.7% are between 70 and 79, there are 28 people or 3.7% who are between 80 and 89,and there are 6 people or 0.8% who are 90 and older.[7]

As of 2000, there were 20 homes with 1 or 2 persons in the household, 116 homes with 3 or 4 persons in the household, and 121 homes with 5 or more persons in the household.[8] As of 2000, there were 262 private households (homes and apartments) in the municipality, and an average of 2.6 persons per household.[6] In 2008 there were 128 single family homes (or 40.1% of the total) out of a total of 319 homes and apartments.[9] There were a total of 3 empty apartments for a 0.9% vacancy rate.[9] As of 2007, the construction rate of new housing units was

9.4 new units per 1000 residents.[6]

In the 2007 federal election the most popular party was the SVP which received 42.09% of the vote. The next three most popular parties were the CVP (25.33%), the SP (10.82%) and the FDP (7.37%). In the federal election, a total of 226 votes were cast, and the voter turnout was 40.7%.[10]

The historical population is given in the following table:[7]

Historical population		
Year	**Pop.**	**±%**
1975	539	—
1980	536	−0.6%
1990	558	+4.1%
2000	685	+22.8%

Heritage sites of national significance

The *Herrain*, a formerly fortified hill, is listed as a Swiss heritage site of national significance.[11]

Economy

As of 2007, Schupfart had an unemployment rate of 1.77%. As of 2005, there were 72 people employed in the primary economic sector and about 28 businesses involved in this sector. 37 people are employed in the secondary sector and there are 9 businesses in this sector. 55 people are employed in the tertiary sector, with 17 businesses in this sector.[6]

In 2000 there were 347 workers who lived in the municipality. Of these, 281 or about 81.0% of the residents worked outside Schupfart while 41 people commuted into the municipality for work. There were a total of 107 jobs (of at least 6 hours per week) in the municipality.[12] Of the working population, 9.7% used public transportation to get to work, and 67.3% used a private car.[6]

Religion

From the 2000 census, 518 or 76.1% were Roman Catholic, while 95 or 14.0% belonged to the Swiss Reformed Church. Of the rest of the population, there were 12 individuals (or about 1.76% of the population) who belonged to the Christian Catholic faith.[8]

Education

In Schupfart about 76.2% of the population (between age 25-64) have completed either non-mandatory upper secondary education or additional higher education (either university or a *Fachhochschule*).[6] Of the school age population (in the 2008/2009 school year), there are 49 students attending primary school in the municipality.[8]

References

[1] Canton of Aargau Statistical Office (https://www.ag.ch/de/dfr/statistik/statistische_daten/oeffentliche_statistik/bevoelkerung/bevoelkerung_1.jsp), MS Excel document – *Bevölkerungsbestand 2012 nach Gemeinde, Geschlecht u. Nationalität (per 31. Dezember 2012)*] (German) accessed 23 July 2013
[2] Arealstatistik Standard - Gemeindedaten nach 4 Hauptbereichen (http://www.bfs.admin.ch/bfs/portal/de/index/themen/02/03/blank/data/gemeindedaten.html)
[3] Swiss Federal Statistical Office-Land Use Statistics (http://www.bfs.admin.ch/bfs/portal/de/index/themen/02/03/blank/data/gemeindedaten.html) 2009 data (German) accessed 25 March 2010
[4] Flags of the World.com (http://www.fotw.us/flags/ch-ag185.html) accessed 28-May-2010
[5] Statistical Department of Canton Aargau -Bereich 01 -Bevölkerung (http://www.ag.ch/staag/) (German) accessed 20 January 2010
[6] Swiss Federal Statistical Office (http://www.bfs.admin.ch/bfs/portal/en/index/regionen/02/key.html) accessed 28-May-2010
[7] Statistical Department of Canton Aargau -Bevölkerungsdaten für den Kanton Aargau und die Gemeinden (Archiv) (http://www.ag.ch/staag/) (German) accessed 20 January 2010
[8] Statistical Department of Canton Aargau - Aargauer Zahlen 2009 (http://www.ag.ch/staag/) (German) accessed 20 January 2010
[9] Statistical Department of Canton Aargau (http://www.ag.ch/staag/) (German) accessed 20 January 2010
[10] Swiss Federal Statistical Office, *Nationalratswahlen 2007: Stärke der Parteien und Wahlbeteiligung, nach Gemeinden/Bezirk/Canton* (http://www.bfs.admin.ch/bfs/portal/de/index/themen/17/02/blank/data/04/03.html) (German) accessed 28 May 2010
[12] Statistical Department of Canton Aargau-Bereich 11 Verkehr und Nachrichtenwesen (http://www.ag.ch/staag/) (German) accessed 21 January 2010

Tabel_Jura

The **Table Jura** is part of the Jura range of mountains in North West Switzerland.

Article Sources and Contributors

Hellikon *Source*: http://en.wikipedia.org/w/index.php?title=Hellikon *Contributors*: 1991gaz, BillC, Bin Cachin', D6, DidiWeidmann, Dr. Blofeld, GoingBatty, Hmains, Ksnow, Kwamikagami, Mandarax, Marek69, Michael Devore, Peter G Werner, Quibik, Rror, The Rambling Man, Tobyc75, Topbanana, 1 anonymous edits

Rheinfelden_District *Source*: http://en.wikipedia.org/w/index.php?title=Rheinfelden_District *Contributors*: Adlei86, BillC, Brewcrewer, D6, Debresser, DidiWeidmann, Docu, Gaius Cornelius, Jeppiz, Markko, Markussep, Matt Gies, MichaelFrey, NoPuzzleStranger, Numbo3, Reza1615, Saltose, Schwyz, That Guy, From That Show!, Tobyc75, Wikibob, ZooFari, 20 anonymous edits

Municipality *Source*: http://en.wikipedia.org/w/index.php?title=Municipality *Contributors*: 12seda78, 16@r, 20060715, 811zssk, 83d40m, A bit iffy, AB-me, AbhiSuryawanshi, AdjustShift, Adnanramin, Adz, Agari, Agathoclea, Ahuskay, Al Ameer son, Alblefter, Aleksandr Grigoryev, Alex Coiro, Alex1011, AlexWangombe, Altenmann, Alvaro, Androl, Ariedartin, Aristitleism, Asim Led, Asocall, Avashnirvana, AxioWiki, Baldhur, Barabum, Bazsimc, Bearcat, Bejnar, Bennylin, Bheru5, BigJolly9, Biglulu, Billinghurst, Bjarki S, Bkonrad, Bob Burkhardt, Bogdan Nagachop, Bogdangiusca, BokicaK, Bomac, Brandon1978, BrentS, BrownHairedGirl, Bryard, BsBsBs, Bspringbett, Buckstars, Bugoslav, CHJL, Cabana, Carabao, Carlosvigopaz, Caroshy, Cassowary, Catrachos, Cetot, ChaChaFut, Chads666, CharlotteWebb, Che090572, Chl, Chris the speller, Christian List, Christopher Sundita, Cncs wikipedia, Conorbrady.ie, Conversion script, Correogsk, Cserlajos, Cubs197, Cyanolinguophile, D6, DAJF, DMahalko, DVdm, Dakilang Isagani, Danny lost, Dantadd, Dave Andrew, Davshul, Dddstone, Den fjättrade ankan, DerHexer, Derek Andrews, Devil Master, Dhum Dhum, Dilic, Djodjo666, Docu, Dom0803, Doublecobbler522, Dreadstar, E Pluribus Anthony, Earl Andrew, Eastlaw, Edcolins, Egil, Eleassar, EoGuy, Epicadam, Eras-mus, Erianna, Espoo, Euchiasmus, Eumolpo, Evlekis, Extra999, Ezhiki, Fadesga, Flyhighplato, Fralambert, Galifardeu, Galoubet, Gazilion, GbrDead, Gene Nygaard, Gerhard51, Gilliam, Glenn, Gwinva, H2ppyme, Haaninjo, Hajor, Halceon, Hamiltha, HappyCamper, Hazhk, Hectorthebat, Hede2000, Herman Downs, Htonl, Hwy43, Ideogram, InimiagoVonKatze, Ionut.sofianu, Iulius, Ivarercool, Jareha, Jaxhere, Jedouellette, JelgavaLV, Jerichochang97, Jerryseinfeld, Jiang, Jim.henderson, Jj137, Jmabel, Jniemenmaa, Joaquin008, Joelr31, JosephIWMolto, Joy, Julius Sahara, Jurema Oliveira, Juro, JustAGal, JustJuthan, Jwillbur, Jóna Þórunn, KJS77, Kafziel, Kalamkaar, Kanags, Knorrepoes, Krystoffer, Kubura, Kuru, Kusma, Kutu su, LA2, LakeHMM, Laurinavicius, Legion23, LeoNomis, Lir, Little sawyer, Lmaltier, Lorenzarius, Lukobe, Lycaon83, MCBastos, MMLeague, Maksim-e, Marcetw, Marek69, Martg76, MauriManya, Mayumashu, Meister, Menchi, Meow, Mercy11, Metropolitan90, Mic, Mieczyslav, Mikko Paananen, Millerjimbones, Motthoangwehuong, Mr. Ajax, Mr. Atom Scania, MrRadioGuy, Mtmelendez, N-edits, NERIUM, Namnguyenvn, Nat Krause, Necrothesp, Nehwyn, Nikola Smolenski, Nipisiquit, Nk, Nricardo, Numbo3, Nurg, Nyttend, OC Ripper, Olessi, Olivier, Oruj, P.Marlow, PBS-AWB, PPP, Pablo.cl, PabloStraub, Pasaquinacity, Patrick, PatrikR, PeacefulKiller, PedroPVZ, Peripitus, Persian Poet Gal, Petr Kopač, Philaweb, Piast93, Picapica, Pinethicket, Po' buster, Postdlf, Puckly, Qyd, Radagast83, Radiojon, Rajankila, Ravichandra71, Remigiu, Retsopllib, Revas, RichardColgate, RingtailedFox, Rjwilmsi, RoMex, Robin Patterson, Robzle, Romanm, Ronline, RoyBoy, Rsocol, RussBlau, S.Örvarr.S, Sam Hocevar, Samudrakula, Sardanaphalus, SchreiberBike, Schwyz, Scott MacLean, Scraimer, Seaphoto, Seav, Selbs101, Sharkb, Sherybatta, Shikuesi3, Skinsmoke, Sluzzelin, Soleado, SomeHuman, Sotonas, Sporkot, Sullu.khan513, Sundar1, Surfo, TBrandley, TGC55, Tabletop, TakuyaMurata, TenIslands, The Transhumanist, TheLeopard, Theshowmecanuck, Thomathekkekara, ThunderingTyphoons!, Tide rolls, Tim Graff, Tkrokli, Tobias Conradi, Torchiest, Turdy wurdy, TurkChan, Twerbrou, UrsoBR, Ushrark, Valentinian, Vcohen, Vera Cruz, VirtualDelight, Vizcarra, Vuo, Wavelength, Wayfarer, WikHead, Wikiwhat?, Wilco499, William Allen Simpson, Woohookitty, Wykebjs, XPTO, Yekrats, Yellaban, Yellowdesk, Yoe, Zellino, Zello, Zero Gravity, ای‌ووپ, 208 anonymous edits

Cantons_of_Switzerland *Source*: http://en.wikipedia.org/w/index.php?title=Cantons_of_Switzerland *Contributors*: 05, 159753, 2Neuenburg, 3riol, A bit iffy, A. Parrot, AP1787, Acntx, Ahuskay, Alarob, Aram33, Argasp, Arthena, Bag, Barticus88, Bearcat, Blanche of King's Lynn, Brammen, Briséis, Brittainia, BrownHairedGirl, CUSENZA Mario, Cassowary, Chanheigeorge, Chris j wood, ChrisfromHouston, Chùn-hiàn, Claude girardin, Closedmouth, Cnilep, Cunado19, D6, DDima, Dbachmann, Dcoetzee, Deacon of Pndapetzim, Denelson83, Djinn112, Docu, Dzordzm, EamonnPKeane, EdwinHJ, El C, Electionworld, Enlil Ninlil, Ericamick, Fan-1967, Filzstift, Frank101, Gazilion, Gestumblindi, Gurch, Haddiscoe, Hashar, Hayden120, Hinote, Hmains, Hroðulf, Hunnjazal, Hvn0413, Indigoeleven, Insilvis, Janburse, JdeJ, Jschwa1, Jusjih, Kahusi, Kerryh r, Kokiri, Korovioff, Lightmouse, Lkmorlan, Longbow4u, Luuva, MadGeographer, Martg76, Mascia Luccova, Master of Puppets, Math Champion, Maxí, Megan1967, MeltBanana, Mistakefinder, MrRadioGuy, Mrt3366, Mschlindwein, Mthwaite, Murtasa, N2e, Nalta, Ngagnebin, Nilfanion, Nohat, Nowhither, Ojw, OwenBlacker, Pascal, Patrick, Paul-L, Piccadilly, Piccolo Modificatore Laborioso, Pne, Postdlf, PsY.cHo, Random user 39849958, Rich Farmbrough, Rjwilmsi, Roke, Rpckvv, Sallygirl1783, Sam Blacketer, Sandstein, Sardanaphalus, Schutz, ScottMainwaring, Sesel, SetRochele, SetRochelewsad, Sherool, SiBr4, Skew-t, Sluzzelin, Smjg, Stephensuleeman, Stevietheman, Tamfang, TenOfAllTrades, Thorsen, Tobias Conradi, Tobyc75, Tonymec, Ultimate Destiny, Uncle Scrooge, Vdiest, West.andrew.g, Woohookitty, Worobiew, Wshun, Zacharie Grossen, Zondor, Žiga, პაატა ვ, 103 anonymous edits

Aargau *Source*: http://en.wikipedia.org/w/index.php?title=Aargau *Contributors*: -- April, 1exec1, 2Neuenburg, Adlei86, Ahoerstemeier, AlexAnnecy, Alfirin, Andre Engels, Aphaia, Aram33, BajanZindy, Baldhur, Bastin, Bob Burkhardt, Bobblewik, Boleyn, ChrisGualtieri, Claude girardin, Conversion script, D6, David Newton, Dbachmann, Dcoetzee, Den fjättrade ankan, Denelson83, Docu, Dr. Blofeld, Dribblingscribe, DwightKingsbury, Editore99, Fastifex, Funandtrvl, Funnyhat, Gilded Lily, Goldfishbutt, Gwern, Habib87, Harleyrules170, Howcheng, Hunnjazal, Ike, Innotata, JM.Beaubourg, Japanese Searobin, Jauhienij, Jbyrde1, Jezerfetnae, KingTT, KirkEN, Klapper, Kokiri, Kummi, Lawrencekhoo, LeoNomis, Lightmouse, Lincher, Longbow4u, MIKHEIL, Macronyx, Magioladitis, Magnum1, Marek69, Matt Gies, Maximaximax, Mgimpel, Mirv, Molly-in-md, Mowsbury, Nancy, Neuenburg, Numbo3, Olivier, Omnipaedista, OwenBlacker, PBS-AWB, Ratcliffe mike, Reisio, Rich Farmbrough, Robert Horning, Rocastelo, SWAdair, Saltose, Scipius, Seanmullan, Shabbs, Sidious1701, Silverhelm, Sluzzelin, Smilo Don, Snori, Snowdog, Speednat, Synthetik, Tobyc75, Trilobite, Vanka5, Varoon Arya, Voyager, Walden69, Wasbeer, WereSpielChequers, Wetman, Wik, Wikibob, Will Beback, Wknight94, Ze miguel, 73 anonymous edits

Basel-Landschaft *Source*: http://en.wikipedia.org/w/index.php?title=Basel-Landschaft *Contributors*: 2Neuenburg, Adlei86, Ahoerstemeier, Al Silonov, Alarob, Andres, Aram33, Awpc, Ben-Zin, BillC, Carolynparrishfan, Claude girardin, D6, Dbachmann, Denelson83, Docu, Ehrenkater, Emerson7, Emperorbma, Eras-mus, Eztoremember, Firefox13, Funandtrvl, Gaius Cornelius, Goldfishbutt, Gologologogg, Grafen, GreatWhiteNortherner, Green Giant, Habib87, Hunnjazal, Johnluisocasio, KirkEN, KnightRider, Kokiri, Kthar, Kummi, Kwamikagami, Lam-ang, Longbow4u, MIKHEIL, Macronyx, Neuenburg, Niteowlneils, Omnipaedista, PsY.cHo, Rjwilmsi, Saxbryn, Skinsmoke, Snowdog, Tobyc75, Trilobite, Trumpnewtq, Unyoyega, Voyager, Walden69, Wik, Wikibob, 51 anonymous edits

Fricktal *Source*: http://en.wikipedia.org/w/index.php?title=Fricktal *Contributors*: BillC, Charles Matthews, D6, Docu, Hmains, Laurascudder, Loveofhistorynut, Ludo29, Matt Gies, NoPuzzleStranger, OwenBlacker, Rich Farmbrough, Sleigh, Tabor, 9 anonymous edits

Obermumpf *Source*: http://en.wikipedia.org/w/index.php?title=Obermumpf *Contributors*: BillC, D6, DidiWeidmann, Dr. Blofeld, Ksnow, Rich Farmbrough, Sole Soul, Tobyc75, 1 anonymous edits

Schupfart *Source*: http://en.wikipedia.org/w/index.php?title=Schupfart *Contributors*: BillC, D6, DidiWeidmann, Dr. Blofeld, Geschichte, Ksnow, New World Man, Rich Farmbrough, Sole Soul, Tobyc75, 1 anonymous edits

Tabel_Jura *Source*: http://en.wikipedia.org/w/index.php?title=Tabel_Jura *Contributors*: Aelfthrytha, Alvestrand, Bin Cachin', H.Grob

Image Sources, Licenses and Contributors

File:Hellikon.jpg *Source*: http://en.wikipedia.org/w/index.php?title=File:Hellikon.jpg *License*: unknown *Contributors*: User:Dimelina

file:Wappen Hellikon AG.svg *Source*: http://en.wikipedia.org/w/index.php?title=File:Wappen_Hellikon_AG.svg *License*: unknown *Contributors*: Sa-se

file:Switzerland location map.svg *Source*: http://en.wikipedia.org/w/index.php?title=File:Switzerland_location_map.svg *License*: unknown *Contributors*: Lukasb1992, Maoman, NordNordWest, Sting, Wikisoft*, 1 anonymous edits

File:Red pog.svg *Source*: http://en.wikipedia.org/w/index.php?title=File:Red_pog.svg *License*: unknown *Contributors*: Anomie

Image:Population District Rheinfelden (CH).svg *Source*: http://en.wikipedia.org/w/index.php?title=File:Population_District_Rheinfelden_(CH).svg *License*: unknown *Contributors*: User:MichaelFrey

Image:Hellikon-blason.png *Source*: http://en.wikipedia.org/w/index.php?title=File:Hellikon-blason.png *License*: unknown *Contributors*: User:Voyager

Image:Kaiseraugst-blason.png *Source*: http://en.wikipedia.org/w/index.php?title=File:Kaiseraugst-blason.png *License*: unknown *Contributors*: User:Voyager

Image:Magden-blason.png *Source*: http://en.wikipedia.org/w/index.php?title=File:Magden-blason.png *License*: unknown *Contributors*: User:Voyager

Image:Moehlin-blason.png *Source*: http://en.wikipedia.org/w/index.php?title=File:Moehlin-blason.png *License*: unknown *Contributors*: User:Voyager

Image:Mumpf-blason.png *Source*: http://en.wikipedia.org/w/index.php?title=File:Mumpf-blason.png *License*: unknown *Contributors*: User:Voyager

Image:Obermumpf-blason.png *Source*: http://en.wikipedia.org/w/index.php?title=File:Obermumpf-blason.png *License*: unknown *Contributors*: User:Voyager

Image:Olsberg-blason.png *Source*: http://en.wikipedia.org/w/index.php?title=File:Olsberg-blason.png *License*: unknown *Contributors*: User:Voyager

Image:Rheinfelden-blason.png *Source*: http://en.wikipedia.org/w/index.php?title=File:Rheinfelden-blason.png *License*: unknown *Contributors*: User:Voyager

Image:Schupfart-blason.png *Source*: http://en.wikipedia.org/w/index.php?title=File:Schupfart-blason.png *License*: unknown *Contributors*: User:Voyager

Image:Stein-blason.png *Source*: http://en.wikipedia.org/w/index.php?title=File:Stein-blason.png *License*: unknown *Contributors*: User:Voyager

Image:Wallbach-blason.png *Source*: http://en.wikipedia.org/w/index.php?title=File:Wallbach-blason.png *License*: unknown *Contributors*: User:Voyager

Image:Wegenstetten-blason.png *Source*: http://en.wikipedia.org/w/index.php?title=File:Wegenstetten-blason.png *License*: unknown *Contributors*: User:Voyager

Image:Zeiningen-blason.png *Source*: http://en.wikipedia.org/w/index.php?title=File:Zeiningen-blason.png *License*: unknown *Contributors*: User:Voyager

Image:Zuzgen-blason.png *Source*: http://en.wikipedia.org/w/index.php?title=File:Zuzgen-blason.png *License*: unknown *Contributors*: User:Voyager

File:Ponce City Hall.JPG *Source*: http://en.wikipedia.org/w/index.php?title=File:Ponce_City_Hall.JPG *License*: unknown *Contributors*: User:Mtmelendez

Image:wikisource-logo.svg *Source*: http://en.wikipedia.org/w/index.php?title=File:Wikisource-logo.svg *License*: unknown *Contributors*: Guillom, INeverCry, Jarekt, MichaelMaggs, NielsF, Rei-artur, Rocket000

File:Suisse cantons.svg *Source*: http://en.wikipedia.org/w/index.php?title=File:Suisse_cantons.svg *License*: unknown *Contributors*: User:Pymouss44

File:Karte 13 Alte Orte.png *Source*: http://en.wikipedia.org/w/index.php?title=File:Karte_13_Alte_Orte.png *License*: unknown *Contributors*: User:Sidonius

Image:Wappen Zürich matt.svg *Source*: http://en.wikipedia.org/w/index.php?title=File:Wappen_Zürich_matt.svg *License*: unknown *Contributors*: Sa-se

Image:Wappen Bern matt.svg *Source*: http://en.wikipedia.org/w/index.php?title=File:Wappen_Bern_matt.svg *License*: unknown *Contributors*: User:Sa-se

Image:Wappen Luzern matt.svg *Source*: http://en.wikipedia.org/w/index.php?title=File:Wappen_Luzern_matt.svg *License*: unknown *Contributors*: Sa-se

Image:Wappen Uri matt.svg *Source*: http://en.wikipedia.org/w/index.php?title=File:Wappen_Uri_matt.svg *License*: unknown *Contributors*: Sa-se

Image:Wappen Obwalden matt.svg *Source*: http://en.wikipedia.org/w/index.php?title=File:Wappen_Obwalden_matt.svg *License*: unknown *Contributors*: Heraldiker Rudolf Münger

Image:Wappen Nidwalden matt.svg *Source*: http://en.wikipedia.org/w/index.php?title=File:Wappen_Nidwalden_matt.svg *License*: unknown *Contributors*: Sa-se

Image:Wappen Glarus matt.svg *Source*: http://en.wikipedia.org/w/index.php?title=File:Wappen_Glarus_matt.svg *License*: unknown *Contributors*: Sa-se

Image:Wappen Zug matt.svg *Source*: http://en.wikipedia.org/w/index.php?title=File:Wappen_Zug_matt.svg *License*: unknown *Contributors*: Sa-se

Image:Wappen Freiburg matt.svg *Source*: http://en.wikipedia.org/w/index.php?title=File:Wappen_Freiburg_matt.svg *License*: unknown *Contributors*: Sa-se

Image:Wappen Solothurn matt.svg *Source*: http://en.wikipedia.org/w/index.php?title=File:Wappen_Solothurn_matt.svg *License*: unknown *Contributors*: Sa-se

Image:Wappen Basel-Stadt matt.svg *Source*: http://en.wikipedia.org/w/index.php?title=File:Wappen_Basel-Stadt_matt.svg *License*: unknown *Contributors*: Sa-se

Image:Coat of arms of Kanton Basel-Landschaft.svg *Source*: http://en.wikipedia.org/w/index.php?title=File:Coat_of_arms_of_Kanton_Basel-Landschaft.svg *License*: unknown *Contributors*: Wappenentwurf: Otto Plattner diese Datei: Jwnabd

Image:Wappen Schaffhausen matt.svg *Source*: http://en.wikipedia.org/w/index.php?title=File:Wappen_Schaffhausen_matt.svg *License*: unknown *Contributors*: (previous version maybe User:Delta-9)

Image:Wappen Appenzell Ausserrhoden matt.svg *Source*: http://en.wikipedia.org/w/index.php?title=File:Wappen_Appenzell_Ausserrhoden_matt.svg *License*: unknown *Contributors*: Sa-se

Image:Wappen Appenzell Innerrhoden matt.svg *Source*: http://en.wikipedia.org/w/index.php?title=File:Wappen_Appenzell_Innerrhoden_matt.svg *License*: unknown *Contributors*: Sa-se

Image:Coat of arms of canton of St. Gallen.svg *Source*: http://en.wikipedia.org/w/index.php?title=File:Coat_of_arms_of_canton_of_St._Gallen.svg *License*: unknown *Contributors*: Wappenentwurf: unbekannt diese Datei: Jwnabd

Image:Wappen Graubünden matt.svg *Source*: http://en.wikipedia.org/w/index.php?title=File:Wappen_Graubünden_matt.svg *License*: unknown *Contributors*: Sa-se

Image:Wappen Aargau matt.svg *Source*: http://en.wikipedia.org/w/index.php?title=File:Wappen_Aargau_matt.svg *License*: unknown *Contributors*: Sa-se

Image:Wappen Thurgau matt.svg *Source*: http://en.wikipedia.org/w/index.php?title=File:Wappen_Thurgau_matt.svg *License*: unknown *Contributors*: Sa-se

Image:Wappen Tessin matt.svg *Source*: http://en.wikipedia.org/w/index.php?title=File:Wappen_Tessin_matt.svg *License*: unknown *Contributors*: Abaddon1337, Ricordisamoa, Sa-se, Skim, 1 anonymous edits

Image:Wappen Waadt matt.svg *Source*: http://en.wikipedia.org/w/index.php?title=File:Wappen_Waadt_matt.svg *License*: unknown *Contributors*: Sa-se

Image:Wappen Wallis matt.svg *Source*: http://en.wikipedia.org/w/index.php?title=File:Wappen_Wallis_matt.svg *License*: unknown *Contributors*: Sa-se

Image:Wappen Neuenburg matt.svg *Source*: http://en.wikipedia.org/w/index.php?title=File:Wappen_Neuenburg_matt.svg *License*: unknown *Contributors*: Sa-se

Image:Wappen Genf matt.svg *Source*: http://en.wikipedia.org/w/index.php?title=File:Wappen_Genf_matt.svg *License*: unknown *Contributors*: Sa-se

Image:Wappen Jura matt.svg *Source*: http://en.wikipedia.org/w/index.php?title=File:Wappen_Jura_matt.svg *License*: unknown *Contributors*: Sa-se

Image:Coat of Arms of Switzerland.svg *Source*: http://en.wikipedia.org/w/index.php?title=File:Coat_of_Arms_of_Switzerland.svg *License*: unknown *Contributors*: User:E Pluribus Anthony, User:Reisio, User:Wiki-vr, User:Zscout370

Image:Karikatur Teilung Basels.jpg *Source*: http://en.wikipedia.org/w/index.php?title=File:Karikatur_Teilung_Basels.jpg *License*: unknown *Contributors*: Alarob, Aliman5040, Infrogmation, Mapmarks, Mattes, Parpan05, Roland zh, Sidonius, Stefan Kühn

File:Aarau Altstadt.jpg *Source*: http://en.wikipedia.org/w/index.php?title=File:Aarau_Altstadt.jpg *License*: unknown *Contributors*: Breeze, Docu, EDUCA33E, FSII, Slomox, Voyager, 1 anonymous edits

File:Berner Aargau.png *Source*: http://en.wikipedia.org/w/index.php?title=File:Berner_Aargau.png *License*: unknown *Contributors*: User:Sidonius

File:Landvogtei Freie Aemter.png *Source*: http://en.wikipedia.org/w/index.php?title=File:Landvogtei_Freie_Aemter.png *License*: unknown *Contributors*: User:Sidonius

File:Grafschaft Baden.png *Source*: http://en.wikipedia.org/w/index.php?title=File:Grafschaft_Baden.png *License*: unknown *Contributors*: User:Sidonius

File:Karte Helvetik 3.png *Source*: http://en.wikipedia.org/w/index.php?title=File:Karte_Helvetik_3.png *License*: unknown *Contributors*: User:Sidonius

File:Lengau AG 5644.jpg *Source*: http://en.wikipedia.org/w/index.php?title=File:Lengau_AG_5644.jpg *License*: unknown *Contributors*: User:Badener

File:Bözberg Lägernblick 8054.jpg *Source*: http://en.wikipedia.org/w/index.php?title=File:Bözberg_Lägernblick_8054.jpg *License*: unknown *Contributors*: Benutzer:BadenerBadener

File:Karte Kanton Aargau Bezirke.png *Source*: http://en.wikipedia.org/w/index.php?title=File:Karte_Kanton_Aargau_Bezirke.png *License*: unknown *Contributors*: Juiced lemon, Tschubby

File:Katholische Stadtkirche mit Sebastianskapelle und Stadtturm II.JPG *Source*: http://en.wikipedia.org/w/index.php?title=File:Katholische_Stadtkirche_mit_Sebastianskapelle_und_Stadtturm_II.JPG *License*: unknown *Contributors*: User:Fanimu

File:KKL Leibstadt.jpg *Source*: http://en.wikipedia.org/w/index.php?title=File:KKL_Leibstadt.jpg *License*: unknown *Contributors*: User:Nawi112

File:Picswiss BL-60-20.jpg *Source*: http://en.wikipedia.org/w/index.php?title=File:Picswiss_BL-60-20.jpg *License*: unknown *Contributors*: Dake, Nicolas Ray

File:Karikatur Teilung Basels.jpg *Source*: http://en.wikipedia.org/w/index.php?title=File:Karikatur_Teilung_Basels.jpg *License*: unknown *Contributors*: Alarob, Aliman5040, Infrogmation, Mapmarks, Mattes, Parpan05, Roland zh, Sidonius, Stefan Kühn

File:Liestal Altstadthaus.jpg *Source*: http://en.wikipedia.org/w/index.php?title=File:Liestal_Altstadthaus.jpg *License*: unknown *Contributors*: User:Ikiwaner

File:Karte Kanton Baselland Bezirke.png *Source*: http://en.wikipedia.org/w/index.php?title=File:Karte_Kanton_Baselland_Bezirke.png *License*: unknown *Contributors*: Juiced lemon, Tschubby

File:Ebm münchestein eingang.jpg *Source*: http://en.wikipedia.org/w/index.php?title=File:Ebm_münchestein_eingang.jpg *License*: unknown *Contributors*: User:Huligan0

File:Schupfart-blason.svg *Source*: http://en.wikipedia.org/w/index.php?title=File:Schupfart-blason.svg *License*: unknown *Contributors*: User:Liftarn

file:Obermumpf-blason.png *Source*: http://en.wikipedia.org/w/index.php?title=File:Obermumpf-blason.png *License*: unknown *Contributors*: User:Voyager

File:Schupfart Kirche.JPG *Source*: http://en.wikipedia.org/w/index.php?title=File:Schupfart_Kirche.JPG *License*: unknown *Contributors*: User:Paebi

file:Schupfart-blason.svg *Source*: http://en.wikipedia.org/w/index.php?title=File:Schupfart-blason.svg *License*: unknown *Contributors*: User:Liftarn

MIX
Papier aus verantwortungsvollen Quellen
Paper from responsible sources
FSC® C105338

Printed by Books on Demand GmbH, Norderstedt / Germany